English for Qualified Dental Employees

Englisch für Zahnmedizinische Fachangestellte

VERLAG EUROPA-LEHRMITTEL · Nourney, Vollmer GmbH & Co. KG

Düsselberger Straße 23 · 42781 Haan-Gruiten

Europa-Nr.: 79572

Autor: Heinz Bendix, 91629 Weihenzell

Ich bedanke mich bei Herrn Dr. med. dent. Hartmut Otto, Ansbach, für seine wertvollen Tipps und Hinweise.

Das vorliegende Buch wurde auf Grundlage der neuen amtlichen Rechtschreibung erstellt.

1. Auflage 2009

Druck 5 4 3 2 1

Alle Drucke derselben Auflage sind parallel einsetzbar, da sie bis auf die Behebung von Druckfehlern untereinander unverändert sind.

ISBN 978-3-8085-7957-2

www.europa-lehrmittel.de
Umschlaggestaltung: idüll, Ulrich Dietzel, 60329 Frankfurt/Main
Satz: Punkt für Punkt GmbH · Mediendesign, 40237 Düsseldorf
Druck: Triltsch Print und digitale Medien GmbH, 97199 Ochsenfurt-Hohestadt

VORWORT

Dieses Lern- und Arbeitsbuch ist speziell für Zahnmedizinische Fachangestellte ***(Qualified Dental Employees)*** im Berufsschulunterricht gedacht.

Es ist für den Selbstunterricht ebenso gut geeignet wie für das neue Unterrichtsprinzip des eigenverantwortlichen, selbstständigen Arbeitens.

Es setzt nur geringfügige Englischkenntnisse voraus.

Sein Schwerpunkt liegt auf situationsbezogenen Dialogen, mit deren Hilfe eine Patientenbegleitung ***(patient attendance)*** auch in englischer Sprache ermöglicht wird. In diesen Dialogen werden grundlegende Strukturen alltäglicher Praxissituationen vermittelt.

Wichtiger als korrekte Grammatik sind dabei die Erweiterung des berufsrelevanten Wortschatzes und das Erlernen von Redewendungen, welche Grundvoraussetzungen für eine flüssige Kommunikation mit Patienten sind und die Sprechfertigkeit fördern.

In den Dialogen wird auf schwer zu merkende, sprachlich und grammatikalisch komplizierte Ausdrucksweisen verzichtet. Die Satzkonstruktionen sind so einfach wie möglich, aber dennoch sprachlich korrekt.

Der Verlag, der Autor und die Berater/innen wünschen viel Erfolg bei der Arbeit mit *English for Qualified Dental Employees/Englisch für Zahnmedizinische Fachangestellte*.

INHALTSVERZEICHNIS

1 A DENTAL APPOINTMENT

Mrs. Nicole Benning, an American lecturer, is presently teaching at a German Fachhochschule. A tooth is causing her trouble, as a filling in one of her back teeth is broken. She needs to see a dentist. One of her colleagues gave her Dr. Frank Zahn's telephone number. His dental surgery is conveniently located near the college. She calls to make an appointment. It is Monday morning. One of Dr. Zahn's dental assistants answers the phone ...

Assistant	(Telefon klingelt) Praxis Dr. Zahn. Sie sprechen mit Katharina Meyer. Guten Morgen.
Mrs. Benning	Hello. Ah ... Do you speak English?
Assistant	A little, yes.
Mrs. Benning	Oh, good. This is Nicole Benning speaking.
Assistant	How can I help you, madam?
Mrs. Benning	I'd like to make an appointment.
Assistant	Have you been here before?
Mrs. Benning	No, I haven't.
Assistant	OK, what's the problem?
Mrs. Benning	The filling in one of my back teeth came out.
Assistant	Oh, I see ... Well, we are very busy this week. Would Friday afternoon be all right for you?
Mrs. Benning	Don't you have anything earlier? This afternoon perhaps?
Assistant	One moment, please. Let me have a look ... No, I'm sorry, we have nothing free this afternoon.
Mrs. Benning	Well, then how about Tuesday afternoon?

UNIT 1

Assistant	Sorry, we are closed on Tuesday afternoon. Wednesday is the earliest we can do. Would 8 o'clock on Wednesday be OK?
Mrs. Benning	8 o'clock is a bit early.
Assistant	Could you come at 10.30 then?
Mrs. Benning	Yes, that's fine.
Assistant	Right ... Could I have your name again please? I didn't catch it.
Mrs. Benning	The name is Benning. Mrs. Nicole Benning.
Assistant	Could you please spell your surname?
Mrs. Benning	You spell it B e n n i n g.
Assistant	OK, Mrs. Benning. See you on Wednesday at 10.30.
Mrs. Benning	Thanks a lot.
Assistant	You are welcome. Have a nice day! Bye!
Mrs. Benning	And you! Bye!

ACTIVITY 1

Important notice! – Wichtiger Hinweis!

Vocabulary work should be done as an essential part of every new lesson. Do it in pairs or in groups. Every student should make his or her personal vocabulary list. Now, find the following words and expressions in the text!

Zahnarzttermin	Lehrbeauftragte, Dozentin	momentan, zurzeit
lehren, unterrichten, dozieren	Kummer, Probleme	eine Füllung
die Backenzähne, die Molaren	(heraus)gebrochen	ein Zahnarzt
dringend	ein(e) Kollege/in	Zahnarztpraxis
günstig gelegen	einen Termin ausmachen	zahnmedizinische Assistentin
Hier spricht ...(Telefon)	Unsere Praxis ist sehr voll ...	etwas früher (Termin)
vielleicht	Ich schau' mal ...	Wir haben nichts frei
Wie wär's mit .../Ginge es ...	Wir haben geschlossen	Ja, das ist gut/Das passt.
Ich habe es nicht verstanden	buchstabieren, schreiben	der Nachname
Bis Donnerstag ...	... um 10 Uhr.	Danke! – Bitte!
Schönen Tag noch!	Ebenfalls!	Tschüss!

ACTIVITY 2

Answer the following questions in English!

1. Who is Mrs. Benning?
2. Why is she staying in Germany?
3. What trouble has she got?
4. Who gave her the dentist's telephone number?
5. Where is Dr. Zahn's dental surgery (auch: dental office) located?
6. Has Mrs. Benning visited Dr. Zahn's surgery (auch: office) before?
7. The dental assistant offers her an appointment for Friday. Is that day OK for Mrs. Benning?
8. Mrs. Benning would like to come on Monday afternoon. Is that possible?
9. What about Tuesday afternoon?
10. Which day would be the earliest the assistant can do?
11. Why does the assistant ask for Mrs. Benning's name again?
12. What would you reply if an English speaking person said "Thank you" to you?

ACTIVITY 3

For Your Information

In Englisch sprechenden Ländern ist eine Zeitangabe wie z. B. 14:50 Uhr i. d. R. noch nicht so üblich wie bei uns, wird aber immer häufiger verwendet. Viele Menschen sagen dort meist noch: "2.50 pm". Die Abkürzung "pm" kommt aus dem Lateinischen und heißt "post meridiem", was soviel bedeutet wie "nach Mittag". Zeitangaben zwischen 12:01 Uhr und 24:00 Uhr werden mit dem Zusatz "pm" versehen. Zeitangaben zwischen 00:01 Uhr morgens und 12:00 Uhr mittags werden mit dem Zusatz "am", d. h., "ante meridiem", also "vor Mittag" versehen. Beispiele: 21:45 Uhr wäre dann 9.45 pm oder auch 21:45 (sprich: twenty-one forty-five). Für zwei Uhr nachmittags (also volle Stunden) kann man z. B.: "2 pm" oder "14:00 hours" (sprich: fourteen hundred hours) sagen.

Beide Möglichkeiten die Uhrzeit anzugeben, können zwar nebeneinander verwendet werden, beachten Sie aber, dass vielen Briten, Iren und Nordamerikanern die 24-Stunden-Zeitrechnung u. U. noch nicht so geläufig ist.

Beispiel:

Für : „Kommen Sie **am** 15. Januar **um** 13:30 Uhr“ kann man sagen:

a) Come on January the 15th, **at** 13.30.

b) Come on January the 15th, **at** 1.30 pm.

Merke: "o'clock" sagt man nur bei vollen Stunden sowie bei 12-Stunden-Zeitrechnung. Beispiel: 18:00 Uhr ist entweder 6.00 o'clock pm oder 18 hundred hours.

Tell the date and the time!

Do it like this: "Please, come on January the 3rd, at 16.30."
"Please, come on January the 3rd, at 4.30 pm."

Now, start telling the time!

03.01./12:30	24.02./08:45	10.03./16:15	17.04./09:30
30.05./14:45	09.06./11:15	31.07./18:00	13.08./12:00
19.09./08:10	10.10./11:10	11.11./11:11	24.12./20:00
06.01./21:45	02.02./23:00	18.03./10:45	20.04./14:15

ACTIVITY 4

Say it in English!

1. Kommen Sie am 23.12. um 14 Uhr.
2. Ihr Termin ist am 14.03. um 9:45 Uhr.
3. Kann ich am 02.07. um 10:30 Uhr kommen?
4. Sie haben Ihren Termin am 19.06. um 11 Uhr.
5. Ich gehe am 7. Februar um 16:15 Uhr zum Zahnarzt.
6. Am 12. Oktober um 11:45 Uhr habe ich einen Termin beim Zahnarzt.
7. Könnte ich am 04.04. um 17 Uhr kommen?
8. Bitte kommen Sie am Montag, den 13. Juli, um 16 Uhr!

ACTIVITY 5

Make sentences!

1. Be/here/before?
2. I/visit/dentist/next Friday.
3. I/have/appointment/dentist/next Tuesday.
4. Can you/come/Thursday/9.30 am?
5. Can/see/doctor/June 6th?
6. I/visit/dentist/last Friday.
7. My/appointment/is/Tuesday/13th.
8. Please/come/Monday/17th/9.30 am.

ACTIVITY 6

Use the information below to make complete sentences.

1. Filling/back tooth/broken.	7. We/busy/this week.
2. The assistant/answer/phone?	8. Surgery/close/Tuesday.
3. I´d like/book/appointment.	9. How/spell/surname?
4. What/be/problem?	10. Surgery/open/Wednesday?
5. Doctor/answer/phone?	11. Who/be/on/phone?
6. I/need/see/dentist.	12. Have/anything/free/Monday?

UNIT 1

ACTIVITY 7

All these statements are false! Please correct them!

1. The filling in one of Mrs. Benning's front teeth is broken.
2. Dr. Zahn's dental surgery/office is far away from the college.
3. Mrs. Benning phones to talk to Dr. Zahn personally.
4. Mrs. Benning visited the surgery last Friday.
5. The dental assistant speaks no English at all.
6. The surgery is not busy this week.
7. There are some appointments free on Monday afternoon.
8. The surgery is open on Tuesday afternoon.
9. Eight o'clock on Wednesday is OK for Mrs. Benning.
10. Mrs. Benning will come on Thursday at 9.30 am.

ACTIVITY 8

Look at the text again and say it in English!

1. Die Patientin ist Lehrbeauftragte/ Dozentin an einer Fachhochschule.
2. Die Füllung in einer ihrer Backenzähne ist herausgebrochen.
3. Eine Kollegin hat mir Ihre Telefonnummer gegeben.
4. Die Zahnarztpraxis liegt in der Nähe der Fachhochschule. Das ist sehr günstig.
5. Stefanie ist eine Zahnmedizinische Fachangestellte.
6. Guten Morgen. Hier spricht Tom Cruise.
7. Ich heiße Bond. James Bond.
8. Was kann ich für Sie tun, Herr Bond?
9. Sind Sie schon einmal hier gewesen?
10. Ich hätte gerne einen Termin.
11. Unsere Praxis ist diese Woche nicht sehr voll.
12. Wäre Ihnen Donnerstagnachmittag recht?
13. Ginge es etwas früher?
14. Vielleicht morgen Nachmittag?
15. Lassen Sie mich nachsehen!
16. Tut mir leid, aber wir haben heute nichts frei.
17. Ginge es Mittwochnachmittag?
18. Tut mir leid, wir haben Mittwochnachmittag geschlossen.
19. Donnerstag wäre das Früheste, was wir anbieten können.
20. Acht Uhr dreißig ist ein bisschen früh.

21. Könnten Sie um 11 Uhr kommen?
22. Ja, das passt.
23. Könnten Sie mir Ihren Namen noch einmal sagen?
24. Ich habe ihn nicht verstanden.
25. Wie schreibt man Ihren Nachnamen?
26. Könnten Sie ihn buchstabieren?
27. Danke! – Bitte!
28. Schönen Tag noch! – Tschüss!
29. Ebenfalls! – Tschüss!

>>> ACTIVITY 9 >>>

Work with a partner. Your partner is a patient. Ask her/him questions. Take turns!

1. May I have your name, please?
2. Could you spell it?
3. Have you been here before?
4. What can I do for you?
5. What's your problem?
6. Could you come on Friday at 9.30?

>>> ACTIVITY 10 >>>

Work with a partner to make a dialogue!

Assistant	Patient
1. Ihr Telefon klingelt. Sie heben ab ...	
	2. Sie sprechen leider nur Englisch. Sie stellen sich vor, nennen Ihr Problem und bitten um einen Termin – möglichst heute noch.
3. Sagen Sie, dass das schwierig ist. Diese Woche ist viel los. Bieten Sie einen Termin an irgendeinem Tag der nächsten Woche an.	
	4. Der angebotene Termin ist viel zu spät für Sie. Bitten Sie um einen früheren. Äußern Sie einen Terminwunsch.
5. Sie sehen im Terminbuch nach. Am gewünschten Tag ist kein Termin mehr frei.	
	6. Äußern Sie einen weiteren Terminwunsch.
7. Ausgerechnet an diesem Tag ist die Praxis geschlossen. Bieten Sie einen anderen Termin an.	
	8. Der Tag würde passen. Sie möchten aber lieber etwas später kommen.
9. Nennen Sie einen späteren Zeitpunkt.	
	10. Sie stimmen jetzt zu.
11. Fragen Sie nochmals nach dem Namen. Sie haben ihn (vorhin) nicht verstanden.	

UNIT 1

12. Sie nennen Ihren Namen nochmals.

13. Sie bitten die Patientin, ihren Namen zu buchstabieren.

14. Sie kommen dieser Bitte nach.

15. Sie bestätigen den Termin.

16. Sie bedanken sich.

17. Sie erwidern den Dank, wünschen noch einen schönen Tag und verabschieden sich.

18. Sie wünschen dies ebenfalls und verabschieden sich.

EXTENDED VOCABULARY LIST	
appointment	Verabredung, Termin; to book or make an appointment
at all	gar nicht, gar kein
back teeth, molars	Backenzähne, Molaren
busy	belebt, verkehrsreich, beschäftigt, voll (Praxis), ausgefüllt
Bye!/Good bye!	Tschüss!/Auf Wiedersehen!, Auf Wiederhören! (Telefon)
catch	fangen (Sachen), einfangen (Krankheit, Virus), verstehen
colleague	Kollege/in
college	Fachhochschule, Akademie, kleinere Universität, Kolleg
convenient/ly	günstig, passend, geeignet
dental assistant **auch: dentist's assistant** **qualified dental employee**	zahnmedizinische Assistentin Zahnarzthelfer/in Zahnmedizinische Fachangestellte
dental office	Zahnarztpraxis (US)
dental surgery	Zahnarztpraxis (UK)
dental treatment	Zahnbehandlung
dentistry	die Zahnmedizin
dialogue	Dialog, Gespräch zwischen zwei Personen
essential	wesentlich, unentbehrlich, unerlässlich

EXTENDED VOCABULARY LIST	
expression	Ausdruck, Redewendung, Ausdrucksweise
extended	erweitert
filling	Füllung
front teeth, incisors	Schneidezähne, Frontzähne
how about	wie wäre es mit ...
locate	etwas finden, lokalisieren
located	er/sie/es liegt, befindet sich
May I ...?	Darf ich ...?, Dürfte ich ...?
need to see	besuchen müssen, aufsuchen müssen
pair/in pairs	Paar/paarweise
part	Bestandteil, Teil, Rolle (Theater)
perhaps	vielleicht
personal	persönlich, personenbezogen
present	anwesend, vorhanden
presently	momentan, zurzeit, bald, in Kürze
reply	erwidern, antworten
some	einige, ein paar, etwas, irgendein
spell	buchstabieren, schreiben
stay	Aufenthalt, sich aufhalten, bleiben, wohnen
surname, family name (UK/US) **second name (US)**	Nachname, Familienname
trouble	beunruhigen, bedrücken, Kummer, Sorgen, Beschwerden
urgently	dringend

An American woman and her husband, on vacation in Germany, went to see a dentist. "I want a tooth pulled, and I don't want Novocaine because we are in a big hurry", the woman said. "Just extract the tooth as quickly as possible." The dentist was impressed. "You are a courageous woman", he said. "Which tooth is it?" The woman turned to her husband and said: "Show him your tooth, darling."

on vacation (US), on holiday (UK) = im Urlaub; to pull a tooth = Zahn ziehen; Novocaine = Novocain (ein Lokalanästhetikum); big hurry = große Eile; impressed = beeindruckt; courageous = mutig; to turn to someone = sich an jemanden wenden)

2 ASKING FOR PERSONAL DETAILS

It is Wednesday morning, 10.15 am. Mrs. Nicole Benning enters Dr. Zahn's dental surgery and walks up to the reception desk. One of the dental assistants is working at the computer ...

Mrs. Benning	Good morning.
Receptionist	Guten Morgen. Wie kann ich Ihnen helfen?
Mrs. Benning	My name is Benning. Nicole Benning. I've got an appointment for 10.30. Ah ... by the way ... can I speak English? My German is still a bit poor, you know ...
Receptionist	No problem, madam.
Mrs. Benning	Oh, that's great.
Receptionist	Before Dr. Zahn can see you I need a few personal details from you.
Mrs. Benning	Sure. What would you like to know?
Receptionist	Could you please put down your first name, your middle name, if you've got one, your last name and your date of birth here on this form? I also need your address here in Germany.
Mrs. Benning	Sure. No problem.
Receptionist	Your nationality is British?
Mrs. Benning	No, I'm American.
Receptionist	Do you have a medical card?
Mrs. Benning	No. I have private dental insurance.
Receptionist	OK. Could you please sign this dental fee agreement? I'm sorry, but we must ask you to pay in cash.

Mrs. Benning Do you also accept credit cards? I've got a Visa Card.

Receptionist I'm afraid, we don't take credit cards but we take EC cards if you've got one.

Mrs. Benning Well, I'll pay cash then. I haven't got an EC card yet. Do you already know what I will have to pay?

Receptionist Sorry, I can't tell you in advance.

Mrs. Benning That's okay.

Receptionist Please, take a seat in our waiting room over there. It will take a few minutes before Dr. Zahn can see you. But it won't take long. Ah ... I almost forgot ... May I ask you to fill in this medical history questionnaire? We usually give it to new patients before they receive dental treatment. And there's one more thing I need from you. Could you also complete this dental hygiene questionnaire and give it back to me when I'll call you? It's important to Dr. Zahn to know some facts about your previous medical history. You can do it while you are sitting in the waiting room.

Mrs. Benning Sure. But ... your questionnaires are all in German!

Receptionist Ah yes. I'll call you in a few minutes and we can do the questionnaires together. Alright?

Mrs. Benning Fine.

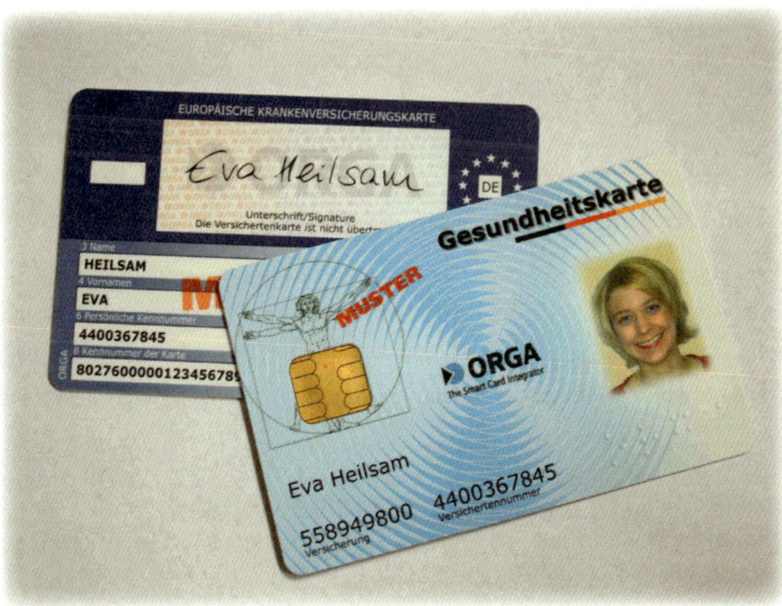

ACTIVITY 1

Find these words and expressions in the text and add them to your vocabulary list!

persönliche Angaben	betreten, eintreten	auf (etwas) zugehen
Anmeldetheke	übrigens ...	schlecht (Sprache)
Das ist toll!, Super!	ein paar, wenige	Klar!, Sicher!
Was möchten Sie gerne?	hin/niederschreiben	zweiter Vorname (US)
Nachname (US)	Formular, Form	Staatsangehörigkeit
KV-Karte, Chipkarte	privat	Zahnversicherung
unterschreiben, unterzeichnen	zahnärztliche Honorarvereinbarung	jemanden bitten
(in) bar	annehmen, akzeptieren	Kreditkarte
nehmen, annehmen	Ich habe sie noch nicht	im Voraus
Platz nehmen	Wartezimmer	dort drüben
Es wird eine Minute dauern	Es wird nicht lange dauern	fast, beinahe
Darf ich Sie bitten ...	ausfüllen (Formular)	Anamnesebogen
üblicherweise	erhalten, bekommen	Zahnbehandlung
Da ist noch etwas ...	vervollständigen, ausfüllen	Zahnhygiene-Fragebogen
zurückgeben	während (Gleichzeitigkeit!)	Tatsachen, Umstände
früher/e, vorhergehend	Zahngesundheit	Ganz in Deutsch
gemeinsam, zusammen	In Ordnung?, Ja?, OK?	Gut!, Schön!

For Your Information

If foreign patients have got a German medical card, you will have to ask them to pay a surgery charge of currently 10 Euros. The surgery charge is due whenever a new quarter (year) starts. However, some patients are exempted from paying the charge. In those cases you have to ask the patient to produce a certificate of exemption issued by their health insurance company.

(foreign = ausländisch; medical card = KV-Karte, Chipkarte; ask = hier: bitten; surgery charge = Praxisgebühr; currently = zurzeit, momentan; it is due = er/sie/es ist fällig; quarter of a year = Vierteljahr; to be exempted from = befreit sein von; in such cases = in diesen/solchen Fällen; to produce, auch: present or show = hier: vorlegen; certificate of exemption = Befreiungsbescheinigung; issued by = ausgestellt von; health insurance company = Krankenversicherung)

ACTIVITY 2

All these statements are false. Correct them!

1. Mrs. Benning's German is very good.
2. Mrs. Benning's last name is Nicole.
3. Mrs. Benning does not have to sign anything.
4. For payment, Mrs. Benning can use her credit card.
5. Dr. Zahn does not accept EC cards for payment.
6. Mrs. Benning will not have to wait.
7. Mrs. Benning has got no problem completing the questionnaires.

ACTIVITY 3

Please, give an answer in English! – Look at the text again!

1. Which personal details does the assistant need from Mrs. Benning?
2. The assistant asks Mrs. Benning to sign something. What is it?
3. Can Mrs. Benning use her credit card for payment?
4. Which two ways of payment are possible at Dr. Zahn's dental surgery?
5. The assistant asks Mrs. Benning to fill in two questionnaires. What questionnaires are they?
6. Why has Mrs. Benning got a problem completing the questionnaires?

American English	British English
to ***book*** an appointment	to ***make*** an appointment
a dental clinic/a dental ***office***	a dental ***surgery***
first name	***Christian*** name
last name	***surname*** or ***family*** name

HINWEIS:

Surgery heißt eigentlich „Chirurgie". Jedoch bezeichnet man damit im britischen Englisch nicht nur die Praxis eines Arztes/Zahnarztes (surgery/dental surgery) sondern auch den eigentlichen Behandlungsraum bzw. das Sprechzimmer. Im amerikanischen Englisch verwendet man für die Praxis eines Arztes oder Zahnarztes dagegen den Begriff *office*.

ACTIVITY 4

Look at the text again and say it in English!

1. Ich benötige ein paar persönliche Angaben von Ihnen.
2. Könnten Sie hier Ihren Vor- und Nachnamen hinschreiben?
3. Könnte ich Ihr Geburtsdatum haben?
4. Welche Staatsangehörigkeit besitzen Sie?
5. Bitte unterschreiben Sie hier!
6. Das ist eine zahnärztliche Honorarvereinbarung.
7. Es tut mir leid, aber wir nehmen keine Kreditkarten an.
8. Haben Sie eine EC-Karte?
9. Darf ich Sie bitten, bar zu zahlen?
10. Nehmen Sie bitte im Wartezimmer Platz.
11. Wo ist es?/Wo befindet es sich?
12. Dort drüben.
13. Ich werde Sie in einigen/ein paar Minuten aufrufen.
14. Es wird nicht lange dauern.
15. Darf ich Sie bitten, diesen Anamnesebogen auszufüllen?
16. Aber – er ist ganz in Deutsch!
17. Könnten Sie auch diesen Fragebogen zur Zahnhygiene ausfüllen?
18. Sie können die Fragebögen ausfüllen, während Sie im Wartezimmer sitzen.

ACTIVITY 5

Work with a partner. Ask her/him questions. Take turns!

1. What is your nationality?
2. What is your surname/last name/family name?
3. Do you have a middle name?
4. Which ways of payment are mentioned in the text?
5. Which way of payment do you prefer when buying things?
6. Can I have your date of birth?
7. Do you know what a Visa Card is?
8. If a foreigner came to your surgery. Would you ask him to pay cash/in cash?
9. Do you accept credit cards at your surgery?
10. Have you ever filled in a medical history/anamnesis form?
11. You want a patient to sit down in the waiting room. How would you politely ask her/him to do so?

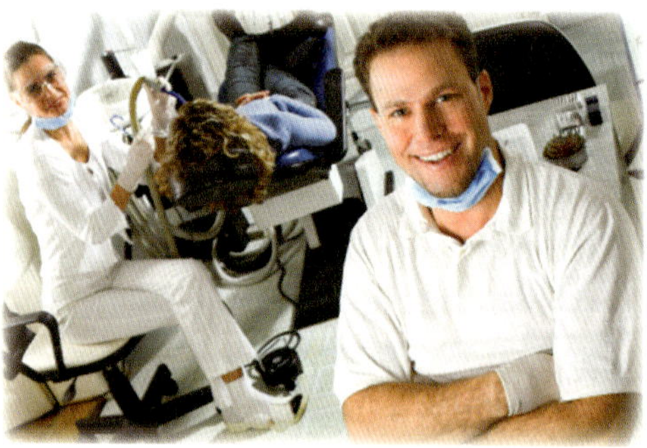

Mother: "Has your tooth stopped hurting yet?"

Son: "I don't know. The dentist kept it."

(to stop hurting = aufhören, weh zu tun; yet = schon; keep = behalten)

>>> ACTIVITY 6 >>>

Tell a patient the way!

Go	round the corner down/along the corridor/hall downstairs upstairs through the door to the ground floor to the first floor to the fourth floor to the top floor

It's	opposite the reception across the road at the end of the corridor/hall

Take	the lift (UK) the elevator (US) the stairs (UK)/the stairway (US)

Turn	left/right

It's	the first door on the left the last door on the right in the basement/on the first floor

Make sentences in English!

Example: "Go upstairs, turn round the corner and then down/along the corridor. It's the third door on the left".

Go on ...

1. Gehen Sie in die oberste Etage, dann durch die Tür und dann um die Ecke. Es ist die zweite Tür rechts.
2. Nehmen Sie den Lift in die dritte Etage, dann nach links und den Korridor entlang. Es ist die letzte Tür rechts.
3. Sie finden den Raum im Kellergeschoss. Nehmen Sie den Lift nach unten, dann durch die Tür und den Gang entlang. Es ist die erste Tür auf der rechten Seite.

4. Gehen Sie um die Ecke, die Treppe runter, dann den Gang entlang und dann nach rechts. Es ist die zweite Tür links.
5. Die Toilette ist am Ende des Ganges.
6. Die Toilette ist gegenüber dem Behandlungsraum.
7. Das Wartezimmer ist auf der gegenüberliegenden Seite des Ganges.

ACTIVITY 7

Make this conversation with a partner in English!

Patient	Assistent
1. Sie betreten die Praxis und grüßen.	
	2. Sie begrüßen den/die Patienten/in und fragen, womit Sie helfen können.
3. Sie bitten darum, Englisch sprechen zu dürfen, weil Ihr Deutsch sehr schlecht ist.	
	4. Das ist kein Problem für Sie.
5. Sie stellen sich vor und sagen, dass Sie um ... Uhr einen Termin haben.	
	6. Sie sehen im Terminbuch nach und bestätigen den Termin. Jedoch bedauern Sie, dass es heute etwas dauern kann.
7. Sie reagieren entsprechend.	
	8. Legen Sie dem/der Patienten/in ein Formular vor und sagen, dass Sie einige persönliche Angaben benötigen. Bitten Sie um Vor- und Nachnamen (evtl. zweiten Vornamen). Außerdem benötigen Sie das Geburtsdatum und die Wohnanschrift.
9. Erwidern Sie etwas.	
	10. Fragen Sie nach der Staatsangehörigkeit.
11. Geben Sie Auskunft.	
	12. Bitten Sie, eine zahnärztliche Honorarvereinbarung zu unterschreiben sowie um Barzahlung.
13. Sie möchten wissen, ob Sie auch per Kreditkarte zahlen können.	
	14. Sie bedauern, dass dies nicht möglich ist. EC-Karten werden angenommen.
15. Erwidern Sie etwas.	
	16. Sie bitten den Patienten/die Patientin, einen Zahnhygiene-Fragebogen sowie einen Anamnesebogen auszufüllen.

17. Sie wollen wissen, weshalb.

18. Erklären Sie, weshalb.

    Die Fragebögen können im Wartezimmer ausgefüllt werden.

    Bitten Sie den Patienten/die Patientin, dort Platz zu nehmen.

19. Sie wissen nicht, wo es sich befindet.

20. Erklären Sie es.

    Die Fragebögen können nach dem Aufrufen an der Anmeldung abgegeben werden.

EXTENDED VOCABULARY LIST	
across **across the road**	auf der gegenüberliegenden Seite auf der anderen Straßenseite
all in German	ganz in Deutsch (gehalten, geschrieben)
anything	irgendetwas
ask	fragen/bitten
basement	Untergeschoss, Kellergeschoss, Tiefparterre
be about	um was es sich handelt Beisp.: What is it about?
by the way ...	übrigens ...
cash/in cash	bar, in bar
corridor (UK)	Flur, Korridor
date of birth	Geburtsdatum
dental health	Zahngesundheit
details	Details, Einzelheiten, Angaben
downstairs **upstairs**	die Treppe hinab, nach unten nach oben
elevator (US) **lift (UK)**	Lift
fact	Einzelheit, Tatsache, Wirklichkeit
fill in, complete	ausfüllen

EXTENDED VOCABULARY LIST	
first name (US) Christian name (UK)	Vorname
foreign	ausländisch
foreigner	Ausländer/in
hall (US)	Flur, Korridor
last name (US)	Familienname
lavatory (UK), toilet (UK) restroom (US)	Toilette
medical history, anamnesis	Anamnese, gesundheitliche Vorgeschichte
mentioned	erwähnt
middle initial	Beisp.: George **W**. Bush, initial = Initiale, Anfangsbuchstabe
middle name	zweiter Vorname (US) Beisp.: George Walker Bush
obliged	verpflichtet
opposite opposite the toilet (UK) opposite the lavatory (UK) opposite the restroom (US)	gegenüber gegenüber der Toilette
payment	Zahlung
personal data	persönliche Daten oder Angaben
personal details	persönliche Daten oder Angaben
personal(ly)	persönlich
polite(ly)	höflich
poor	arm, schlecht, unzureichend
possible	möglich
prefer	vorziehen, bevorzugen
previous	früher/e, vorhergehend
questionnaire	Fragebogen
receive	erhalten, empfangen, bekommen
reception	Anmeldung, Empfang
receptionist	Empfangschef oder -dame

EXTENDED VOCABULARY LIST	
seat	Sitz, Sitzplatz
Take a seat!	Nehmen Sie Platz! (Auf keinen Fall: "Sit down!" sagen!!)
sign	unterzeichnen, unterschreiben, Zeichen
stairs (UK) **stairway (US)**	die Treppe
surgery (UK), office (US)	Arztpraxis
surgery (US/UK)	die Chirurgie
take	nehmen, annehmen, akzeptieren, ertragen
top floor	oberste Etage
way	Weg, Weise, Art und Weise

A little boy was taken to the dentist. He had a cavity that would have to be filled. "Now, young man", asked the dentist, "what kind of filling would you like for your tooth?" "Chocolate, please", replied the youngster.

(to take to = hier: jemanden irgendwo hinbringen; cavity = Loch, Kavität; what kind of = was für eine (Art von); reply = erwidern; youngster = Junge, Jugendlicher)

3 A MEDICAL HISTORY QUESTIONNAIRE

While Dr. Zahn is treating a patient, his assistant has some time left to help Mrs. Benning complete the medical history questionnaire. She asks Mrs. Benning to come along to a small room next to the reception ...

Assistant Well, Mrs. Benning. Dr. Zahn thinks it is necessary that his patients complete this form here. It's a medical history questionnaire. Afterwards, we would like you to fill in another questionnaire which we also give to our patients. The second one contains questions about your personal oral hygiene. We can do that at your next appointment. Now, let's start with some questions about your previous medical history ...

OK, Mrs. Benning. My first question is: Do you know the approximate date of your last visit to a dentist?

Mrs. Benning Oh, let me think ... Hm ... About two years ago, perhaps?

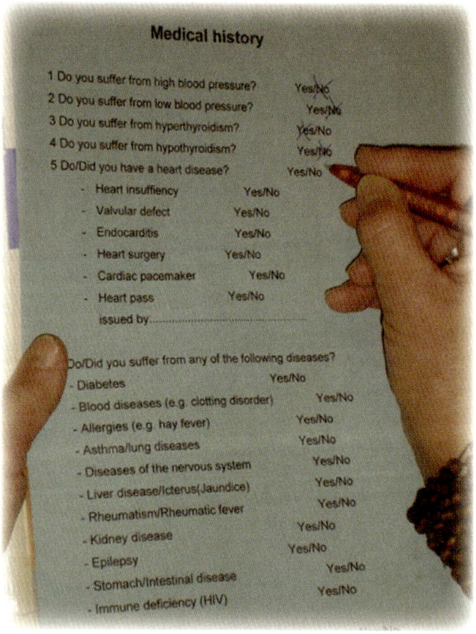

Assistant What was the reason for your last visit to a dentist?

Mrs. Benning I had two wisdom teeth pulled.

Assistant When did you last have a dental X-ray taken?

Mrs. Benning Hm ... I just can't remember exactly. Five years ago, perhaps?

Assistant Are you taking antibiotics, anticoagulants or medicine for high blood pressure?

Mrs. Benning Excuse me, what are "anticoagulants"?

Assistant Anticoagulants are blood thinners.

Mrs. Benning Ah, I see. No, I'm not taking any of those medicines.

Assistant Are you taking any other medication at the present time?

Mrs. Benning No, I don't.

Assistant	Are you allergic to certain medicines?
Mrs. Benning	Well, I'm allergic to pollen and animal hair but I do not know whether I'm allergic to certain medicines.
Assistant	Do you wish to get local anaesthesia?
Mrs. Benning	Oh yes. I hate having pains.
Assistant	OK. That's all. I'm going to attach this form to your patient's file. Thanks a lot for giving me your time, Mrs. Benning.
Mrs. Benning	No problem. You're welcome.
Assistant	Dr. Zahn is now ready to see you, Mrs. Benning. Could you come along to the exam room? I'll show you the way ...

>>> ACTIVITY 1 >>>

Find these words or expressions in the text and add them to your vocabulary list!

Anamnese	Fragebogen	etwas Zeit übrig haben
vervollständigen, ausfüllen	jem. bitten, etwas zu tun	mitkommen
gleich neben, gleich daneben	nötig, notwendig	Formular, Vordruck
danach, hinterher	ausfüllen	der zweite (z. B. Vordruck)
er enthält	eine Frage nach etwas	persönlich
Zahnhygiene	einige Fragen zu	früher, vorhergehend
das ungefähre Datum	ich ließ sie ziehen (hier: Weisheitszähne)	Zahnröntgen
vor fünf Jahren	Antibiotika	gerinnungshemmende Mittel, Antikoagulanzien
blutverdünnende Mittel	Medikamente	gegenwärtig, zurzeit
allergisch (reagieren) auf	bestimmte Arzneimittel	Pollen
Tierhaare	lokale, örtliche Betäubung	beilegen, beifügen (als Anlage)
Akte, Büroordner	Danke! – Bitte!	er ist soweit
mitkommen	Untersuchungszimmer exam(ination) room	weisen, zeigen

ACTIVITY 2

Please, give an answer to these questions on the text!

1. What are Dr. Zahn's patients usually asked to complete before he starts his treatment?
2. What kind of questions does the second questionnaire contain?
3. What is the approximate date of Mrs. Benning's last visit to a dentist?
4. What was the reason for her visit?
5. Did Mrs. Benning ever have dental X-rays taken? If yes, when was that?
6. Can she remember the date exactly?
7. Is she taking anticoagulants, antibiotics or medicine for high blood pressure?
8. Is she allergic to certain medication/medicines?
9. Does she suffer from any other allergies?
10. Why does she wish to get local anaesthesia if necessary?
11. What will the assistant do with the medical history questionnaire?
12. Which room does the assistant show her the way to?

ACTIVITY 3

Please translate these sentences into English!

1. Ich habe etwas Zeit übrig, um Ihnen zu helfen.
2. Könnten Sie bitte diesen Anamnese-(frage)bogen ausfüllen?
3. Könnten Sie danach (noch) einen weiteren Fragebogen ausfüllen?
4. Der zweite enthält Fragen zu Ihrer Zahnhygiene.
5. Sie können ihn bei Ihrem nächsten Termin ausfüllen.
6. Darf ich Ihnen einige Fragen zu Ihrer medizinischen Vorgeschichte stellen?
7. Wissen Sie (noch) den ungefähren Zeitpunkt Ihres letzten Zahnarztbesuchs?
8. Was war der Grund Ihres letzten Zahnarztbesuchs?
9. Haben Sie noch Ihre Weisheitszähne?
10. Wurden bei Ihnen schon einmal die Zähne geröntgt?
11. Nehmen Sie zurzeit Antibiotika oder blutverdünnende/gerinnungshemmende Medikamente?
12. Leiden Sie an hohem Blutdruck?
13. Nehmen Sie zurzeit irgendwelche anderen Medikamente?
14. Sind Sie allergisch auf bestimmte Medikamente?
15. Wünschen Sie eine örtliche Betäubung?
16. Ich werde diesen Fragebogen Ihrer Patientenakte beilegen.
17. Danke, dass Sie sich für mich Zeit genommen haben.
18. Oh, nichts zu danken!/Oh, bitteschön.
19. Könnten Sie bitte mit mir mitkommen?

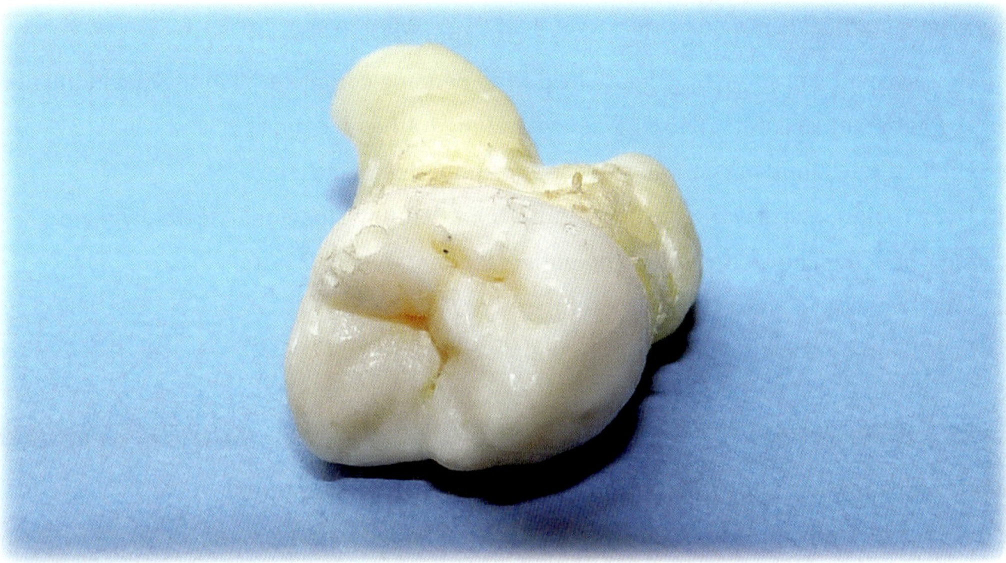

A Wisdom Tooth

ACTIVITY 4

Check these statements! If they are false, please correct them!

1. Anticoagulants are medicines for high blood pressure.
2. A medical history questionnaire contains questions about a patient's dental hygiene.
3. Mrs. Benning can still remember the exact date of her last dental X-ray.
4. Mrs. Benning had two front teeth pulled at her last visit to a dentist.
5. Mrs. Benning suffers from a pollen allergy.
6. The assistant will throw away Mrs. Benning's medical history questionnaire.
7. The second questionnaire contains questions about a patient's previous medical history.

ACTIVITY 5

Work with a partner. You make the following statements. Your partner checks them and says either "yes" or "no". Take turns. Remember that in English, you should use question tags like "Yes, she has", "Yes, he does", "Yes, he will" or "No, it isn't", "No, she hasn't", "No she doesn't", "No, she won't" and so on.

Now, let's start!

1. The assistant has some time left to help Mrs. Benning complete/fill in the questionnaire.
2. The assistant asks Mrs. Benning to come along to the reception.
3. The first questionnaire contains questions about dental hygiene.
4. The assistant starts with questions about Mrs. Benning's personal dental hygiene.
5. Mrs. Benning knows the exact date of her last dental X-ray.
6. Mrs. Benning will have a wisdom tooth pulled.
7. Mrs. Benning is taking medicines at the present time.

>>> ACTIVITY 6 >>>

Work with a partner. Your partner answers your questions. Take turns!

1. Is the dentist in your dental surgery interested in the patients' previous medical history?
2. Do you give any questionnaires to your patients? If yes, what kind of?
3. Did you ever have a dental X-ray taken? If yes, when was that?
4. Do you still have all your wisdom teeth? If yes, are they causing you any trouble?
5. Have you ever taken any antibiotics? If yes, can you remember why?
6. Do you know what an "anticoagulant" is?
7. Do anticoagulants kill bacteria?
8. Does anyone of your family suffer from high blood pressure?
9. Are you allergic to certain medicines?
10. Why do some people wish to get local anaesthesia?
11. What would you reply if a patient said "Thank you" to you?
12. If you want a patient to follow you to the exam(ination) room. How would you tell him to do so?

>>> ACTIVITY 7 >>>

Make this conversation with a partner!

Assistent	Patient
1. Rufen Sie den Patienten/die Patientin auf und bitten Sie ihn/sie, mitzukommen. Sie möchten dabei helfen, den Anamnesebogen auszufüllen. Bitten Sie den Patienten/die Patientin, in einem kleinen Raum Platz zu nehmen. Beginnen Sie mit einer Frage nach seinem/ihrem letzten Zahnarztbesuch.	
	2. Geben Sie eine Antwort.

3. Fragen Sie nach dem Grund des letzten Zahnarztbesuches.

4. Geben Sie eine Antwort.

5. Sie möchten wissen, ob beim Patienten/ bei der Patientin schon einmal die Zähne geröntgt wurden.

6. Geben Sie eine Antwort.

7. Fragen Sie, welche Medikamente zurzeit eingenommen werden. Geben Sie Beispiele.

8. Geben Sie eine Antwort.

9. Fragen Sie, ob der Patient/die Patientin an irgendwelchen Allergien leidet.

10. Geben Sie eine Antwort.

11. Erkundigen Sie sich nach eventuellen Medikamentenallergien

12. Geben Sie eine Auskunft.

13. Fragen Sie, ob lokale Betäubung erwünscht ist.

14. Geben Sie eine Antwort.

15. Sagen Sie dem Patienten/der Patientin, dass dies alles war und Sie den Anamnesebogen der Patientenakte beilegen. Bedanken Sie sich.

16. Erwidern Sie den Dank.

EXTENDED VOCABULARY LIST	
allergic to	allergisch auf/gegen
allergy	Allergie; he suffers from an allergy = er ist Allergiker
antibiotic	Antibiotikum
anticoagulant	gerinnungshemmendes Mittel, Antikoagulans
approximate/ly	ungefähr, annähernd, etwa
as, because	da, weil
attach	beifügen, anheften, befestigen
attachment	Anlage, Beilage, Befestigung, Zusatzvorrichtung

EXTENDED VOCABULARY LIST	
coagulate	gerinnen
coagulation	Gerinnung (z. B. des Blutes)
come along	mitkommen
complete, fill in	vervollständigen, ausfüllen
contain	enthalten, beinhalten
container	Behälter
date	Datum, Zeitangabe, Zeitpunkt, Termin
exact/ly	genau, exakt, pünktlich, gewissenhaft, passend
file	Aktenordner, Akte, Datei, ablegen, abheften
form	Formular, Vordruck, Form, Gestalt, Umgangsform, Schema
be left **von: leave/left/left**	übrig sein
local	lokal, örtlich
local anaesthesia	lokale Betäubung, Lokalanästhesie
local anaesthetic	Lokalanästhetikum
located	(örtlich) gelegen
pain	Schmerz, Pein; (aber: toothache, headache; ache = weh)
perhaps	eventuell, vielleicht
pollen	Blütenstaub, Pollen
present	gegenwärtig, augenblicklich, anwesend, vorhanden
receive	empfangen (Menschen), erhalten (Post)
reception	Anmeldung, Rezeption, Empfang
receptionist	Rezeptionist/in, Empfangsdame, Empfangschef
whether	ob
wisdom	Klugheit, Weisheit
X-rays	Röntgenstrahlen
X-ray	das Röntgenbild, das Röntgen

One day, an American couple walks into a German dentist's office and asks how much it will cost to extract two wisdom teeth. "Two hundred Euros", the dentist says.

"That's too much", the man says. "Isn't there a cheaper way?"

"Well," the dentist says, if I don't use an anaesthetic, I can reduce the price to 100 Euros." "That's still too expensive," the man says. "Okay," says the dentist. "If I don't use an anaesthetic and simply rip out the teeth with a pair of pliers, I can reduce the price to 50 Euros."

"No," says the man, "it's still too much." "Well," says the dentist, "if I let one of my assistants do it, I can reduce the price to 25 Euros."

"Great," says the man, "book my wife for next Tuesday!"

(couple = Ehepaar; extract = extrahieren, ziehen; wisdom teeth = Weisheitszähne; cheaper way = billigere Art und Weise; anaesthetic = Narkose(mittel); reduce = reduzieren; simply = einfach; rip out = herausreißen; a pair of pliers = eine Zange; book my wife = tragen Sie mir einen Termin für meine Frau ein)

4 DENTAL TREATMENT

Mrs. Benning follows Dr. Zahn's dental assistant to the treatment room. Dr. Zahn's chair-side assistant Barbara, asks Mrs. Benning to take a seat on the dental chair ...

Assistant So, please take a seat. I'll just put this bib on you. Dr. Zahn will be with you in a minute, Mrs. Benning.
Ah ... could you open your mouth? OK, that's good. You said a filling in one of your back teeth is broken? Ah ... yes. There's a filling here that is broken. The other molars seem to be OK. But, there's a lot of calculus around your back teeth.

Mrs. Benning Calculus? What's that?

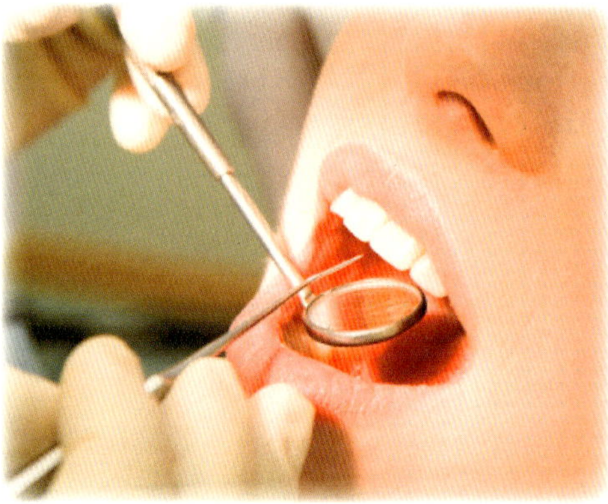

Assistant That's tartar.

Mrs. Benning I see ...

Assistant And there is a lot of swelling of your gums.

Mrs. Benning Yes, I know.

Assistant Do your gums bleed when you brush your teeth?

Mrs. Benning Sometimes, yes.

Assistant Your front teeth look a bit yellow ...

Mrs. Benning Yes, I know. Can you make them nice and white again?

Assistant Our dental hygienist will do her best. She will remove all the tartar and the plaque that is the yellow crust on your teeth, as well.

Mrs. Benning Will you do that today?

Assistant I'm afraid this will take some more time. We'll make another appointment for you for next week. Oh ... there's a small black spot on one of your incisors!

Mrs. Benning Oh, should it bother me?

Assistant You should know that this is caries!

Mrs. Benning Caries?

Assistant Yes, that's tooth decay. That's how it starts. We must remove it. You'll need a filling afterwards.

UNIT 4

Mrs. Benning A filling in one of my front teeth? Will it show much?

Assistant Don't worry, we normally use fillings which do not show. So, you won't see it.

Mrs. Benning Fine.

Assistant If you don't mind, our dental hygienist could give you some advice on how to improve your dental care. We want you to keep your healthy smile for the rest of your life!
You will see, after Dr. Zahn's treatment, your smile will brighten up a room when you walk into it ...

>>> ACTIVITY 1 >>>

Find these words and expressions in the text and add them to your vocabulary list!

zahnärztliche Behandlung	Behandlungsraum	(Behandlungs)stuhl-assistentin
Untersuchungsstuhl	Latz, Lätzchen	Zahnstein (2 Begriffe möglich)
Tatsächlich?, Echt?	Schwellung	Zahnfleisch
Schneidezähne	Ich tue mein Bestes	wegputzen
beseitigen, entfernen	Zahnbelag, Plaque	Ablagerung, Kruste
Fleck, Stelle, Tupfen	Schneidezähne (latein.)	stören, plagen, quälen
Zahnverfall, Zersetzung, Fäule	hinterher, danach, später	hier: auffallen
Füllung	Wenn es Ihnen nichts ausmacht ...	Zahnhygienikerin
einige Ratschläge, Tipps geben	verbessern	Zahnprophylaxe, Vorsorge
erhalten, behalten	gesundes Lächeln	zum Leuchten bringen

A woman phones her dentist as she received a huge bill from him. "I'm shocked!" she complains. "This is three times what you normally charge."

"Yes, I know," the dentist says. "But you cried so loud that two other patients got scared and left the surgery."

(receive = erhalten; huge bill = eine riesige Rechnung; complain = sich beschweren; three times = dreimal; charge = berechnen; get scared = Angst bekommen; leave/left/left = gehen, abfahren, verlassen)

ACTIVITY 2

Look at the text again and give an answer!

1. Mrs. Benning has dental problems. Which are they?
2. What is Dr. Zahn's dental hygienist going to do about Mrs. Benning´s teeth?
3. What did the hygienist find on one of her front teeth?
4. Mrs. Benning is worried about getting a filling in her incisor. Why?
5. What kind of fillings does Dr. Zahn usually use?

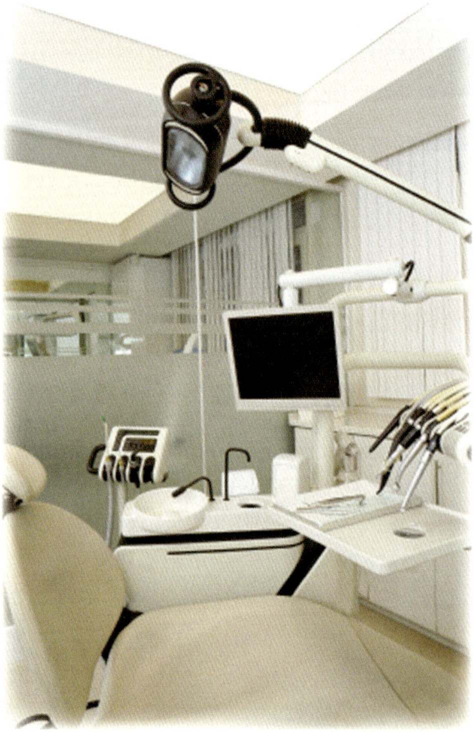

ACTIVITY 3

Please, translate these sentences!

1. Nehmen Sie bitte auf dem Untersuchungsstuhl Platz.
2. Dr. Zahn wird gleich kommen.
3. Könnten Sie Ihren Mund öffnen?
4. OK, so ist's gut.
5. Die Füllung in einem Ihrer Backenzähne ist herausgebrochen.
6. Die anderen Backenzähne scheinen in Ordnung zu sein.
7. Sie haben viel Zahnstein an Ihren Backenzähnen.
8. Ihr Zahnfleisch ist stark geschwollen.
9. Ihre Schneidezähne schauen ein bisschen gelb aus.
10. Der gelbe Belag auf Ihren Zähnen ist Plaque.
11. Unsere Zahnhygienikerin wird den Zahnstein und den Zahnbelag entfernen.
12. Das wird eine Weile dauern./Das wird etwas Zeit erfordern.
13. Dieser schwarze Punkt hier ist Karies!

14. Dr. Zahn wird ihn entfernen.
15. Danach werden Sie eine Füllung bekommen.
16. Wir verwenden nur Füllungen, die nicht auffallen.
17. Unsere Dental-Hygienikerin wird Ihnen ein paar Tipps/Ratschläge geben, wie Sie Ihre Prophylaxe (noch) verbessern können.
18. Wir wollen (doch) Ihr gesundes Lächeln erhalten!

>>> ACTIVITY 4 >>>

Work with a partner. Ask her/him questions!

1. Have you got any fillings in your molars or incisors?
2. Do your gums bleed sometimes?
3. What is plaque?
4. What is the first sign of caries?
5. Do you know another expression for caries?
6. Could you give a patient advice on how to improve his or her dental care? If yes, what advice would you give him or her?

>>> ACTIVITY 5 >>>

Make this conversation with a partner!

Chairside Assistant	Patient
1. Bitten Sie den Patienten/die Patientin, auf dem Behandlungsstuhl Platz zu nehmen. Binden Sie ihm/ihr einen Latz um. Bitten Sie ihn/sie, den Mund weit zu öffnen. Sehen Sie sich die Zähne an und beschreiben Sie, was Sie sehen.	
	2. Ihre Zähne bluten beim Putzen.
3. Sie stellen etwas Zahnstein und Zahnbelag fest. Sagen Sie, dass Sie Beides entfernen werden, nachdem sich der Zahnarzt/die Zahnärztin Ihre Patientin angesehen hat.	
	4. Fragen Sie, ob dies heute noch geschieht.
5. Bejahen Sie es. Sagen Sie, dass es vermutlich etwas dauern wird.	
	6. Erwidern Sie, dass es kein Problem wäre, wenn es nicht allzu lange dauert.

7. Sie finden eine kariöse Stelle an der Rückseite eines Schneidezahns. Weisen Sie die Patientin darauf hin, dass diese vom Zahnarzt/von der Zahnärztin entfernt werden muss.

8. Fragen Sie, ob der Zahnarzt/die Zahnärztin bohren muss und ob es wehtun wird.

9. Beantworten Sie die Frage.

10. Fragen Sie, ob Sie danach eine Füllung bekommen.

11. Bejahen Sie dies.

12. Fragen Sie, ob die Füllung stark auffallen wird.

13. Verneinen Sie dies. Der Zahnarzt verwendet ausschließlich Füllungen, die nicht auffallen

Patient: "Doctor, I'm very nervous.
You know, this is my first extraction!"

Dentist: "Don't worry, it's my first extraction too."

EXTENDED VOCABULARY LIST	
bib	Schürzenlatz, Lätzchen
chairside	die Seite neben einem Stuhl
chairside assistant	auf Stuhlassistenz spezialisierte Helferin
drill	bohren, Bohrer
front	Vorderfront, Uferpromenade, Fassade, Stirn
be in front of someone	vor jemandem sein
in front	vorne
at the front	an der Spitze
frontgarden	Vorgarten
seem	scheinen
He seems younger than he is.	Er wirkt jünger, als er ist.
Things aren't always what they seem.	Vieles ist anders, als es aussieht.
It only seems like it.	Das kommt einem nur so vor.

EXTENDED VOCABULARY LIST	
show	zeigen, auffallen
swell up	anschwellen, sich aufblähen, sich ausdehnen, zunehmen
swell/swelled/swollen **auch: swell/swelled/swelled**	schwellen/schwoll/geschwollen
swelling	Schwellung, Geschwulst, Beule
syringe	(Injektions-)Spritze
anaesthetic syringe	Narkosespritze
dental syringe	Injektionsspritze
to syringe	ausspülen, spritzen, injizieren
disposable syringe	Einmalspritze, Einwegspritze, Wegwerfspritze
hypodermic syringe	Injektionsspritze zur subkutanen Injektion

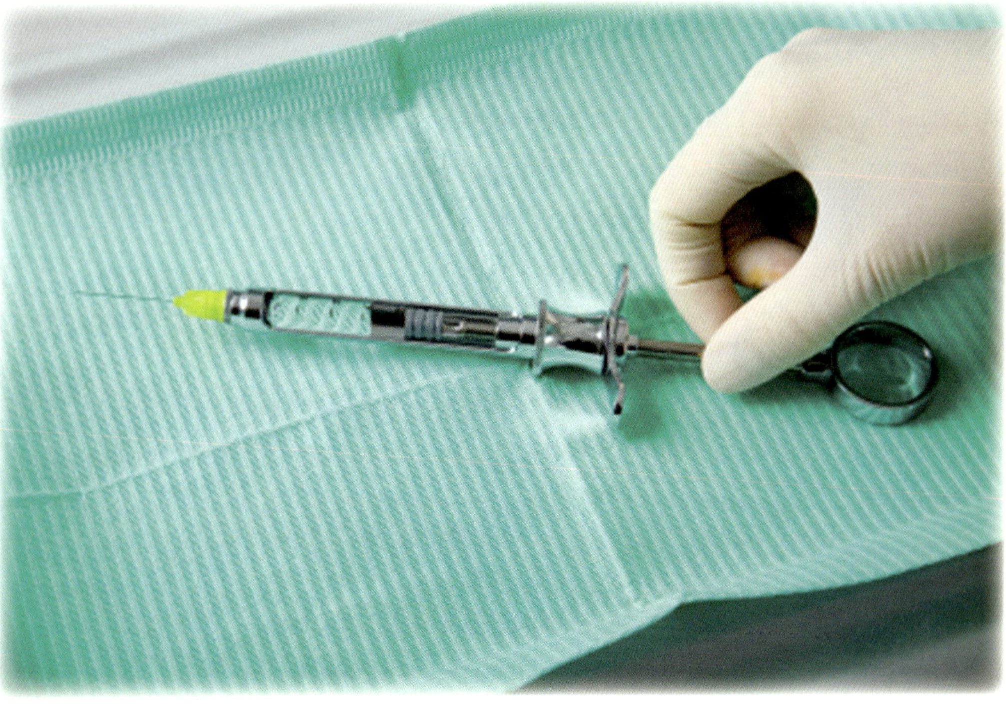

An Anaesthetic Syringe

5 DENTAL HYGIENE

Mrs. Benning has got another appointment with Dr. Zahn's dental hygienist one week after her first visit to his dental surgery. Before the dental hygienist starts the cleaning, she asks Mrs. Benning a few questions about her personal dental hygiene. Together they complete a dental hygiene questionnaire ...

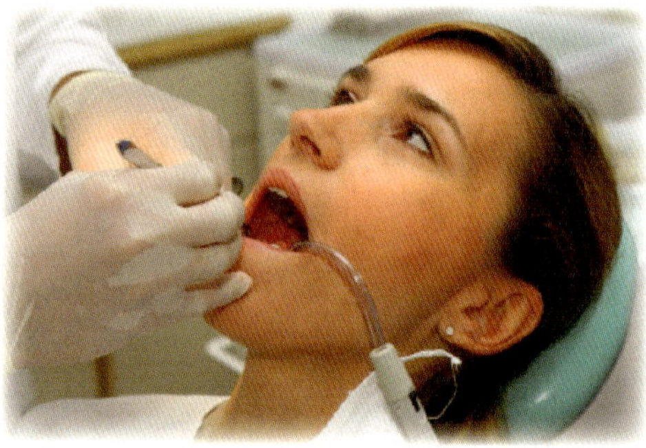

Hygienist	Before we start, I'd like to ask you a few questions about your personal dental hygiene. Which dental care products do you use?
Mrs. Benning	Well, I use an electric toothbrush with a small head and a mouthwash.
Hygienist	Do you use dental floss?
Mrs. Benning	No. Not really.
Hygienist	Have you ever used dental gum?
Mrs. Benning	No, I haven't.
Hygienist	How often do you change the head of your electric toothbrush?
Mrs. Benningg	Oh ... about every 4–6 month.
Hygienist	How often do you brush your teeth?
Mrs. Benning	Once a day. In the morning.
Hygienist	How often do you visit a dentist?
Mrs. Benning	Oh, not very often. Only when I have a problem with my teeth. I know, I'm a bit careless.
Hygienist	What could be a reason for you to visit a dentist?
Mrs. Benning	Well, a reason could be a broken filling or to stop toothache.
Hygienist	Looking at your answers, I would advise you to visit a dentist not only if your teeth cause you trouble but for regular check-ups and also for dental hygiene.
Mrs. Benning	How often should I visit a dentist?
Hygienist	Once a year, or even better every 6 month.
Mrs. Benning	I see ...

Hygienist You should change the head of your electric toothbrush more often. I'd recommend you to do that every 2–3 months.

Mrs. Benning OK.

Hygienist I'd advise you to clean your teeth morning and evening. You need not do it after every meal. However, a sugar-free dental gum after meals might be helpful.

Mrs. Benning Why should I use dental gum after meals?

Hygienist Well, dental gum reduces the acidity in your mouth. It's the acid in your mouth that attacks your gums and the dental enamel and causes tooth decay.

Mrs. Benning The acid causes tooth decay? How come?

Hygienist The acid destroys the enamel on the outside of your teeth and can form cavities in your teeth.

Mrs. Benning I always thought it was the sugar in the food, which causes caries ...

Hygienist It's not the sugar, it's the acid. The bacteria in your mouth turn the sugar into an aggressive acid. That's why it is a good thing to use dental gum after having sweet snacks, to reduce the acidity in the mouth. A mouthwash may also be helpful as it kills the bacteria in your mouth. However, mouthwashes often contain powerful disinfectants. Those disinfectants are dangerous if you swallow them and some of them can cause discoloration of the teeth when used daily.

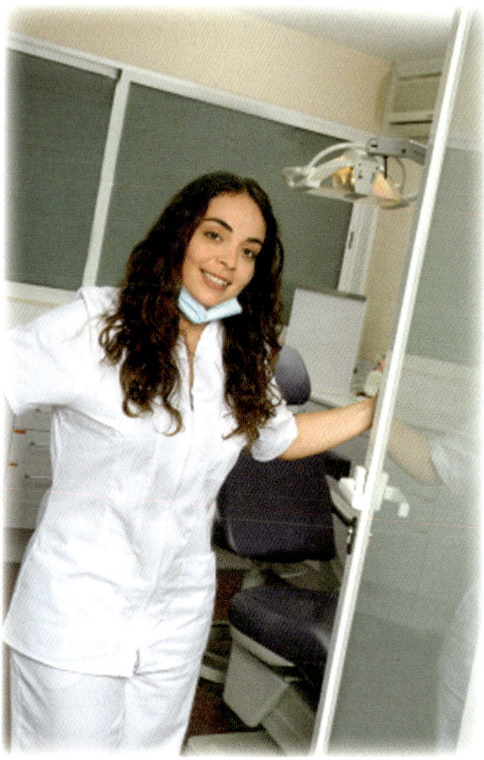

Mrs. Benning Oh, I didn't know that ... Thanks for telling me all this!

Hygienist Oh, you're welcome. That's my job anyway ...

ACTIVITY 1

Find these words or expressions in the text and add them to your vocabulary list!

persönliche Zahnhygiene	Zahnpflegeprodukte	ein kleiner (Zahnbürsten-)Kopf
eine Mundspülung	Zahnseide verwenden	eigentlich nicht
Zahnkaugummi	den (Zahnbürsten-)Kopf wechseln	einmal pro Tag
nachlässig, sorglos, unvorsichtig	ein Grund für	aufhalten, beenden
Wenn ich mir die Antworten (so) ansehe ...	raten, empfehlen	Probleme bereiten
regelmäßig	noch besser	Ah, ja ...
empfehlen	morgens und abends	Mahlzeit, Essen
zuckerfrei	es könnte hilfreich sein	den Säuregehalt vermindern
das Zahnfleisch angreifen	Karies, Zahnfäule,	Wieso?, Wie kommt das?
den Zahnschmelz zerstören	an der Außenseite	Löcher, bzw. Zahnhöhlen bilden
den Zucker in Säure verwandeln	süße Zwischenmahlzeiten	Bakterien abtöten
enthalten	starke Desinfektionsmittel	verschlucken, schlucken
die Zähne verfärben	Bitte schön!, Keine Ursache!	ohnehin, sowieso

ACTIVITY 2

Look at the text and give an answer!

1. What questions does the hygienist ask before she starts the cleaning?
2. Which dental care/dental hygiene products does Mrs. Benning use?
3. How often does Mrs. Benning change the head of her electric toothbrush?
4. How often does she brush her teeth?
5. How often does she visit a dentist?
6. Does Mrs. Benning take good care of her teeth?
7. The hygienist gives Mrs. Benning some advice on her dental care/hygiene. What does she advise her to do?
8. What effect does dental gum have?
9. How does the acid affect the teeth?
10. What effects and disadvantages do mouthwashes have?

ACTIVITY 3

Look at the text again and translate these sentences!

1. Bevor ich mit der Reinigung beginne, möchte ich Ihnen ein paar Fragen zu Ihrer persönlichen Zahnhygiene stellen.
2. Welche Zahnpflegeprodukte verwenden Sie?
3. Verwenden Sie Zahnseide?
4. Wie oft putzen Sie Ihre Zähne?
5. Sie sollten den Kopf Ihrer elektrischen Zahnbürste öfter wechseln.
6. Wie oft gehen Sie zum Zahnarzt?
7. Sie sollten Ihre Zähne morgens und abends putzen.
8. Was war der Grund Ihres letzten Zahnarztbesuchs?
9. Ich würde Ihnen raten, den Zahnarzt für regelmäßige Untersuchungen und zur Zahnhygiene aufzusuchen.
10. Sie sollten einmal jährlich, oder besser noch, alle sechs Monate zum Zahnarzt gehen.
11. Ein zuckerfreier Zahnkaugummi ist sehr hilfreich.
12. Ein Zahnkaugummi vermindert den Säuregehalt im Mund.
13. Es ist die Säure, die den Zahnschmelz und das Zahnfleisch angreift.
14. Die Säure kann Löcher in den Zähnen bilden.
15. Eine Mundspülung ist hilfreich, weil sie die Bakterien abtötet, welche Karies verursachen.
16. Mundspülungen enthalten sehr starke Desinfektionsmittel.
17. Bei täglicher Anwendung können einige von ihnen Verfärbungen an den Zähnen verursachen.

ACTIVITY 4

Work with a partner and ask her or him questions!

1. Which dental care products do you use?
2. How often do you change your toothbrush/the head of your electric toothbrush?
3. How often do you brush your teeth?
4. How often do you visit a dentist?
5. Do you have any dental problems at the present time?
6. Do you go for regular dental check-ups?
7. How does the acid in your mouth arise?
8. The acid in your mouth is harmful. Why?
9. What can you do to reduce harmful acid in your mouth?

ACTIVITY 5

Make this conversation with a partner!

Assistent	Patient
1. Bitten Sie die Patientin, ihr ein paar Fragen zu ihrer persönlichen Zahnhygiene stellen zu dürfen.	
	2. Sie bejahen die Bitte.
3. Fragen Sie sie, welche Zahnpflegeprodukte sie verwendet.	
	4. Zählen Sie einige auf.
5. Fragen Sie, wie oft und weshalb sie zu einem Zahnarzt geht.	
	6. Geben Sie eine Antwort.
7. Beurteilen Sie die Zahnpflege der Patientin und machen Sie weitere Vorschläge, falls sich die Zahnpflege noch verbessern lässt. Weisen Sie auf die Schädlichkeit von Zuckergenuss hin. Erklären Sie auch Vor- und Nachteile von Mundspülungen.	
	8. Das wussten Sie bisher noch nicht. Bedanken Sie sich für die Informationen.
9. Erwidern Sie den Dank der Patientin.	

UNIT 5

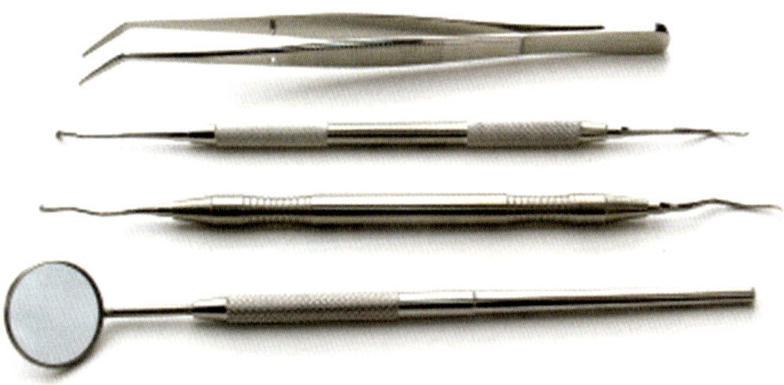

EXTENDED VOCABULARY LIST	
arise (from/out of)	entstehen aus, sich ergeben, auftauchen (Probleme)
careless	sorglos, gedankenlos, achtlos, nachlässig
carefree	ohne Sorgen
careful	sorgfältig, sorgsam, vorsichtig
present	zurzeit, gegenwärtig, vorhanden, momentan

For Your Information

About Toothpaste ...

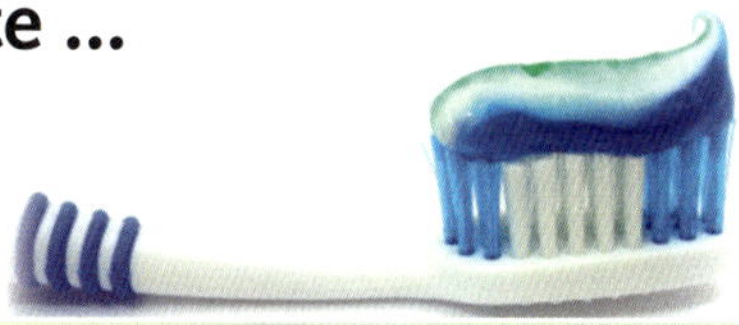

Do you know what ingredients (= Inhaltsstoffe) there are in your toothpaste? You don't? Have a look!

Allantoin	Allantoin	wirkt hautberuhigend und wundheilend, beugt Zahnfleischröte und Zahnfleischbluten vor
Aminfluoride	Aminfluorid	kationisches Tensid; lagert sich an Zahnoberflächen an; bildet Schutzfilm; hemmt Entmineralisierung
Aqua	lat. Bezeichnung für Wasser	dient zur Feuchthaltung der Zahncreme
Sorbitol	Sorbitol	süß schmeckender Zuckeraustauschstoff
Aroma	Aromastoff	verantwortlich für den Geschmack
Calcium carbonate	Kalziumkarbonat, kohlensaurer Kalk, Schlämmkreide	Hilfs- und Trägerstoff in Zahncremes, wirkt abrasiv, als Putzkörper enthalten
Calcium peroxide	Kalzium Superoxid	in umweltfreundlichen Bleich-, Reinigungs- und Desinfektionsmitteln enthalten, bleicht und reinigt durch Sauerstoff
Chamomilla extract	Kamillenextrakt	wirkt hautberuhigend, heilend
Gum	Milchsaft, bzw. Harz vom Gummibaum (Akazie)	Verdickungsmittel
Hydrated silica	wasserhaltige Kieselerde	milde Putzkörper, abrasiv (abschleifend), zur Entfernung von Plaque
Hydrochloric acid	Salzsäure	verhindert Infektionen, wirkt antibakteriell und antimikrobiell
Hydrogenated castor oil	hydriertes, d.h. gehärtetes Rizinusöl	Emulgierhilfsmittel; dient zur Herstellung von Cremes und Seifen
Lauryl glucoside	Laurylglukosid	hautverträgliches, reinigendes Zuckertensid

Limonene	Limonene	Hydrokarbonat, starker Orangengeschmack, künstlicher Geschmackstoff
Potassium chloride	Kaliumchlorid, Kalisalz	wirkt schmerzhemmend, in hohen Dosen toxisch, in Zahncremes für empfindliche Zähne
Potassium hydroxide	Kaliumhydroxid, Ätzkali	senkt den Säuregehalt im Mund
Potassium nitrate	Kaliumnitrat, Kalisalpeter	wirkt antibakteriell
Propylene glycol	Propylenglykol	antimikrobielles Lösungsmittel in Hygieneartikeln
Sodium C14–16 olefin sulfonate	Natrium-Olefinsulfonat	reinigendes und schaumbildendes Tensid
Sodium chloride	Natriumchlorid, Natriumsalz	hemmt Schmerzempfindung, in Zahncremes für empfindliche Zähne enthalten
Sodium fluoride	Natriumfluorid	Zweck: siehe Aminfluorid; bei Überdosierung Zahnverfärbungen und Zahnstrukturstörungen möglich
Sodium gluconate	Natriumglukonat	stellt Säure-Basen-Gleichgewicht wieder her
Sodium lauryl sulfate	Natriumlaurylsulfat	schaumbildendes Tensid
Sodium methylparaben	Natriummethylparaben	hemmt die Entwicklung von Mikroorganismen, Konservierungsstoff
Sodium nitrate	Natriumnitrat, Natronsalpeter	wirkt antibakteriell
Sodium sulfate	Natriumsulfat	reguliert den pH-Wert im Mundraum
Strontium chloride	Strontiumchlorid	reduziert die Schmerzempfindlichkeit der Zahnhälse
Tetrapotassium pyrophosphate	Tetrakaliumsalz	dient der Erhöhung der Phosphatkonzentration im Mundraum
Titanium dioxide	Titandioxid	natürliches Mineral, wirkt als Antioxidant, UVA/UVB Strahlen reflektierend, in allen Sunblockern
Xanthan	Xanthan	Verdickungsmittel

Xylitol	Xylit, Xylitol	Zuckeralkohol; hemmt Bakterienwachstum und somit Plaque; remineralisiert Zähne

Important notice!

Never swallow toothpaste when cleaning your teeth!

There was a Bible verse hanging in a dental office:

Psalms 81 verse 10: "Open your mouth wide and I will fill it."

(verse = Vers; fill = füllen, plombieren, stopfen; to fill a mouth = im übertragenen Sinne auch: Hunger stillen/Durst löschen)

6 PROFESSIONAL DENTAL CLEANING

Dr. Zahn replaced Mrs. Benning's broken filling. He had given her an injection with a local anaesthetic before he treated the cavity in one of her front teeth because he had to do some drilling. Mrs. Benning hates having pain. Now, he is finished. His dental hygienist enters the treatment room and takes care of Mrs. Benning ...

Hygienist Hi, Mrs. Benning. How are you today?

Mrs. Benning I'm fine. Thanks.

Hygienist OK. Could you open up wide? ... That's good. I'll start with removing the tartar from your molars and incisors. Afterwards I'll remove the plaque. Could you turn your head towards me? If you need to rinse your mouth or spit, please let me know.

Mrs. Benning What do you mean by plaque?

Hygienist Plaque is the yellowish crust on and in between your teeth.

Ah ... is this incisor here at the front a dental crown ...?

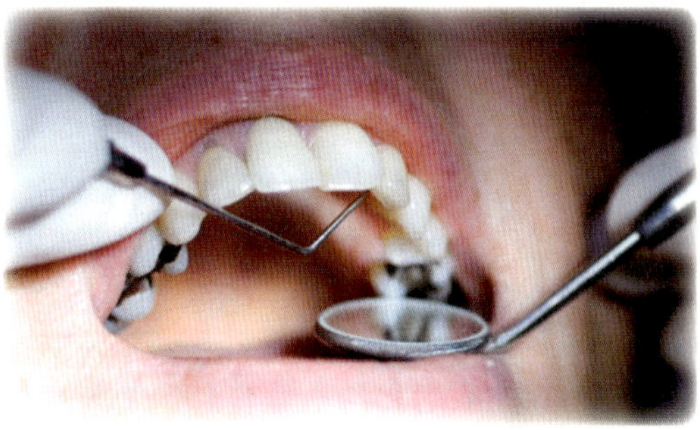

Mrs. Benning Yes, I broke this tooth when I was a teenager.

Hygienist There's quite a lot of tartar around your crown!

Mrs. Benning Can I see it in the mirror?

Hygienist Sure. Here you are. (She hands her a small mirror.)

Mrs. Benning Oh, you're right.

About 20 minutes later

Hygienist OK. Now, I'll polish your teeth, apply a high fluoride gel and then we are done.

Mrs. Benning May I have a look?

Hygienist Is it OK for you? (She hands her the mirror once again.)

Mrs. Benning My teeth definitely look clean now. But, they don't look as white as I have expected it.

Hygienist Well, that's the natural colour of your teeth.

Mrs. Benning I expected my teeth to become really white ...

Hygienist Well, what I've done just now, is what we call a "professional dental cleaning". However, what you are talking about is "aesthetic", or in other words, cosmetic dental treatment. That's something very different. All those people having super-white teeth have had cosmetic dental treatment, you know. Professional dental cleaning makes your teeth clean but does not change the natural colour of your teeth.

Mrs. Benning Why not?

Hygienist You see, professional dental cleaning means that we only clean surfaces. This is something mechanical. We use an ultrasound dental scaler and a double-curved scaler to remove tartar and plaque from the surface of your teeth. Only chemicals are able to penetrate dental enamel and bleach stains and discolorations within the enamel. Only bleaching can change the colour of the enamel.

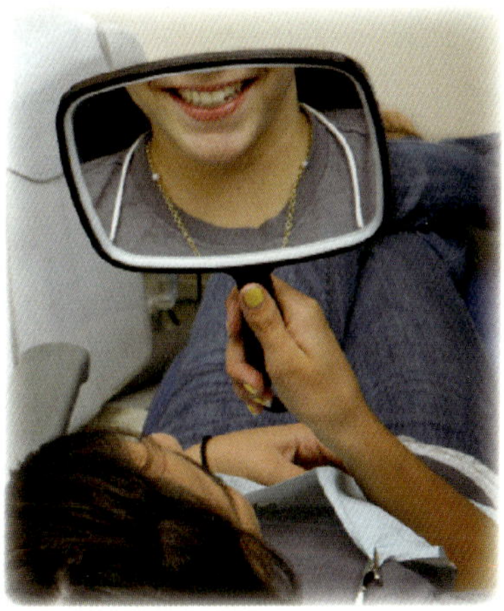

Mrs. Benning Ah, I see. Could you give me more information on bleaching at my next dental appointment?

Hygienist Sure I can!

ACTIVITY 1

Find these words or expressions in the text and add them to your vocabulary list!

etwas ersetzen	eine Spritze/Injektion geben	Lokalanästhetikum
ein wenig bohren müssen	er ist fertig (mit der Arbeit)	Behandlungsraum
sich um etwas kümmern	mit den Backenzähnen anfangen	den Kopf drehen
in meine Richtung, auf mich zu	spülen und spucken müssen	gelbliche Ablagerung
vorne	eine (Zahn-)Krone	ziemlich viel

im Spiegel	Klar!, Sicher!	Hier, bitte!
geben, aushändigen, überreichen	richtig, ordentlich, geeignet	das Polieren
dann sind wir fertig	ganz bestimmt, definitiv	die natürliche Farbe
erwarten	kosmetisch	ästhetisch
Wissen Sie, ...	es heißt, es bedeutet	Oberflächen reinigen
etwas Mechanisches	Ultraschallgerät zum Entfernen von Zahnstein	doppelt-gebogene Kürette
den Zahnschmelz durchdringen	Flecken und Verfärbungen bleichen	Na klar kann ich das!

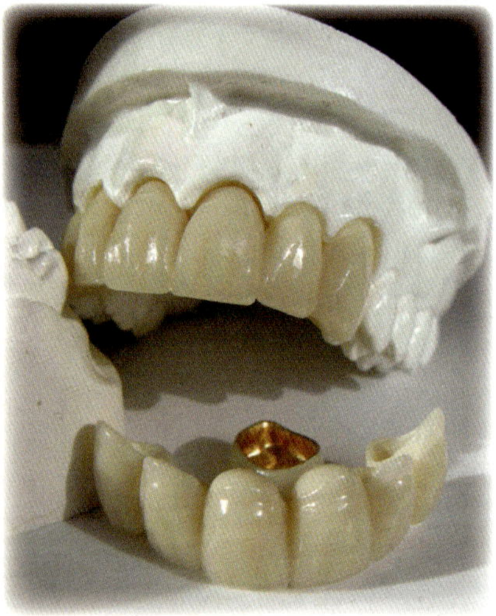

Dental Crowns

ACTIVITY 2

Look at the text and give an answer!

1. What did Dr. Zahn give Mrs. Benning before he replaced her filling?
2. Which teeth did the hygienist start the cleaning with?
3. What did she remove first and what did she do after that?
4. What is "plaque"?
5. One of Mrs. Benning's teeth is a dental crown. How come?
6. What did the hygienist find around her crown?
7. How did Mrs. Benning react when she saw her teeth in the mirror after the cleaning was done?
8. Does professional dental cleaning change the natural colour of the teeth?
9. How can you get "super-white" teeth?
10. What instruments do dentists and hygienists use for cleaning the teeth?

ACTIVITY 3

Look at the text again and translate these sentences!

1. Ich werde etwas bohren müssen.
2. Wünschen Sie eine örtliche Betäubung?
3. Meine Assistentin wird sich danach um Ihre Zähne kümmern.
4. Können Sie (Ihren Mund) weit aufmachen?
5. Ich fange mit Ihren Backenzähnen an.
6. Zuerst werde ich den Zahnstein entfernen, dann den Zahnbelag.
7. Plaque ist die gelbliche Ablagerung an und zwischen den Zähnen.
8. Zwei meiner Schneidezähne sind Kronen.
9. Es befindet sich viel Zahnstein um Ihre Kronen herum.
10. Zahnstein und Zahnbelag sind jetzt entfernt.
11. Ich poliere (noch) Ihre Zähne und dann sind wir fertig.
12. Meine Zähne sehen aber nicht so weiß aus wie ich es erwartet habe!
13. Das ist Ihre natürliche Zahnfarbe!
14. Ich möchte gerne wirklich weiße Zähne.
15. „Professionelle Zahnreinigung" bedeutet nicht, dass die natürliche Zahnfarbe verändert werden kann.
16. Falls Sie super-weiße Zähne wünschen, benötigen Sie eine kosmetische Zahnbehandlung.
17. Nur chemische Mittel können den Zahnschmelz durchdringen und die Zahnfarbe verändern.

ACTIVITY 4

Ask your partner a few questions! Take turns!

1. Do you have any fillings? If yes, where do you have them?
2. If you were a patient and the dentist had to do some drilling, would you like to have local anaesthesia?
3. Have you ever removed tartar in your dental office/surgery?
4. Do you know what "plaque" is?
5. Do you have any gaps between your teeth?
6. Have you ever had professional dental cleaning?
7. Some people's teeth still look yellowish after a professional dental cleaning. How come?
8. What would you recommend a patient who would like to have really white teeth?
9. Which instruments do you use for professional dental cleaning?

Double Curved Dental Scalers

ACTIVITY 5

Make this dialogue with a partner!

Assistent	Patient
1. Bitten Sie die Patientin, ihren Mund weit aufzumachen. Sagen Sie ihr, mit welchen Zähnen Sie beginnen werden. Sie beginnen zuerst mit der Zahnstein-, dann mit der Plaque-Entfernung.	
	2. Sie wissen nicht genau, was die Assistentin mit „Plaque" meint.
3. Erklären Sie es. Sie stellen fest, dass ein Schneidezahn eine Krone hat. Fragen Sie, was passiert ist.	
	4. Erzählen Sie, wie es dazu kam.
5. Sie stellen viel Zahnstein an den Zähnen fest.	
	6. Sie möchten es selbst sehen und bitten um einen Spiegel.
7. Sie geben ihr einen kleinen Spiegel.	
	8. Sie bestätigen die Feststellung.
9. Sie sagen der Patientin, dass Sie ihre Zähne abschließend polieren werden und dann fertig sind.	
	10. Nach Abschluss der Behandlung bitten Sie nochmals um den Spiegel.
11. Geben Sie der Patientin den Spiegel.	
	12. Sie sind etwas enttäuscht, weil Ihre Zähne nicht gerade super-weiß geworden sind.
13. Sagen Sie ihr, dass das etwas mit der natürlichen Zahnfarbe zu tun hat.	
	14. Fragen Sie, wie man richtig weiße Zähne bekommen kann.
15. Erklären Sie der Patientin, was für eine Behandlung erforderlich wäre, um die natürliche Zahnfarbe zu verändern und wirklich weiße Zähne zu erzielen.	
	16. Bedanken Sie sich für die Informationen.
17. Erwidern Sie den Dank.	

EXTENDED VOCABULARY LIST	
aesthetic	ästhetisch
apply	auftragen, anwenden
bleach stains and discolourations	Flecken und Verfärbungen bleichen
crown **dental crown**	Krone Zahnkrone
crust	Kruste, Schorf, Ablagerung
definite/ly	bestimmt, klar, deutlich, unmissverständlich
done **to be done**	getan fertig sein, etwas erledigt haben, erschöpft oder kaputt sein
expect	erwarten
finished **Are you finished?**	fertig sein Sind Sie fertig?
gap	Lücke
hand	die Hand, überreichen
How come?	Wieso? Wie kam es dazu? Woher kommt das?
Is it OK for you?	Passt es Ihnen so?
May I have a look?	Darf ich mal sehen?
mechanical	mechanisch
mirror	Spiegel
natural colour	Naturfarbe, natürliche Farbe
penetrate	eindringen, durchdringen
polish	polieren, Politur
quite a lot	ziemlich viel
rinse **rinse down** **rinse out**	spülen, die Spülung, die Tönung (Haarfarbe) abspülen ausspülen
scaler **dental scaler** **ultrasound dental scaler, also:** **ultrasonic scaler** **double-curved dental scaler**	Scaler Zahnsteinschaber Ultraschallgerät zur Entfernung von Zahnstein und Zahnbelag doppelt-gebogener Scaler zur Zahnstein-entfernung

EXTENDED VOCABULARY LIST	
surface	Oberfläche, Belag
take care	sich kümmern um
What do you mean by ...?	Was meinen Sie mit ...?
within	innerhalb, im Inneren von
yellowish	gelblich

A dentist is begging the patient: "Could you help me? Could you give out a few of your loudest, most painful screams?"

Patient: "Why? It isn't all that painful this time."

Dentist: "There are still so many patients in the waiting room. I want them to leave. I don't want to miss the 4 o'clock football game!"

(beg = bitten, betteln; to give out = ausstoßen, von sich geben; most painful screams = schlimmste Schmerzensschreie; all that painful = nicht so schmerzhaft; this time = diesmal; miss = versäumen, vermissen; game = Spiel)

7 COSMETIC DENTISTRY

Mrs. Benning is still unhappy about the appearance of her teeth. They are clean now but not really "super-white" as she had expected it. She is convinced that a "Hollywood" smile would make her look more impressive and she would feel more confident when talking to people. She decides to make another dental appointment. On the day of her appointment she enters the dental surgery to get some information on "bleaching".

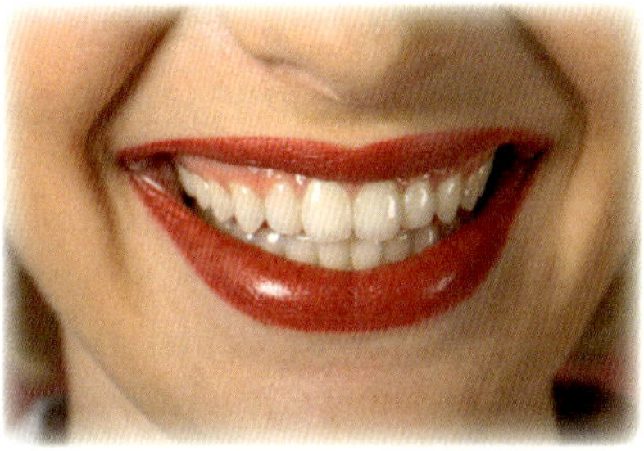

Assistant Hello, Mrs. Benning.

Mrs. Benning Hello, Barbara.

Assistant Did you know that even the old Romans tried to bleach discoloured teeth? They used the urine of Portuguese donkeys. And it worked!

Mrs. Benning How very funny, but I do hope you use different methods nowadays ...

Assistant Don't worry, Mrs. Benning. No donkey's urine anymore. We use chemicals instead.

Mrs. Benning Someone told me that drugstores and Internet pharmacies offer bleaching sets which are very cheap compared to "In-office-bleaching".

Assistant I cannot recommend you to use "over the counter" products. Only a dentist or dental hygienist can do that professionally. There is a great danger of harming your gums or damaging your teeth for the rest of your life. You should know that the chemicals used for bleaching are aggressive, can cause burns, irritate the skin and the gums and can even "kill" dental nerves. They require careful handling.

Mrs. Benning What chemicals do you use to bleach the teeth?

Assistant We use carbamide peroxide gel. The active agent in carbamide peroxide is hydrogen peroxide. The bleaching obtained is proportional to the length of time the peroxide is applied to the teeth and the percentage of hydrogen peroxide used.

Mrs. Benning How long will the procedure take?

Assistant It takes either 8 hours, one hour or just 10 minutes dependent on the method you wish to have applied.

Mrs. Benning Which bleaching methods do you use in your dental office?

Assistant Well, basically there are three methods:

The most commonly used method is "home bleaching". We make a custom-made whitening tray for your teeth which you fill with a 3% hydrogen peroxide gel that we prepare for you. You fill the tray before going to sleep and wear it overnight for about 8 hours. The treatment is finished after about 7 to 10 nights.

A faster method is "bleaching in the waiting room". The procedure is the same. However, we use a 12% hydrogen peroxide gel and you wear the tray while sitting in the waiting room. It takes no longer than one hour.

The fastest method is "bleaching in the dental chair". It only takes about 10 minutes. However, we have to use a 35% hydrogen peroxide gel. This concentration is most aggressive to the gums. That is why it is very important to protect the gingiva with a barrier membrane, called "dental dam".

The whole procedure sometimes has to be repeated several times. It depends on how discoloured the teeth are.

Mrs. Benning Are there any side-effects or disadvantages?

Assistant As far as I know there aren't. The enamel, the dentin, the pulp and the surfaces of fillings or false teeth will not be harmed. However, during treatment, the teeth may become sensitive to cold or heat, because of dehydration of the teeth. To avoid dehydration, you should not apply the peroxide gel every night, but only every second or third night or in the case of "in-office-bleaching", it is advisable to wait 2–3 days before applying the gel again.

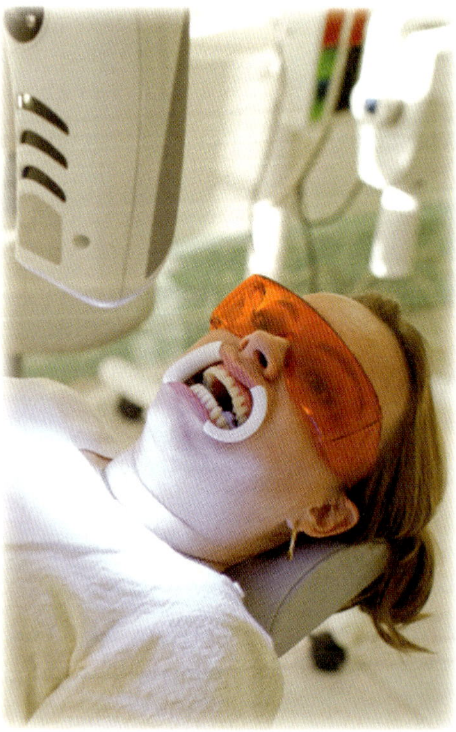

Mrs. Benning How does bleaching work?

Assistant Well, the oxygen can easily penetrate the enamel, where it oxidises, that is, bleaches stains and discolourations.

Mrs. Benning What about "whitening toothpastes"?

Assistant Well, contrary to what many people think, whitening toothpastes do not increase the whiteness of the teeth. However, what they effectively do is removing plaque and stains caused by coffee, tea or tobacco. Such toothpastes contain harsh abrasives. Over time, those harsh abrasives destroy tooth enamel making the teeth appear yellow and causing the teeth to become sensitive to cold, hot, sweet or salty foods and liquids.

Mrs. Benning What toothpastes should I look for then?

Assistant Look for toothpastes with fluoride and those which are minimally abrasive. Also look for toothpastes which contain calcium peroxide and xylitol. They non-abrasively remove stains on the teeth.

>>> ACTIVITY 1 >>>

Find these words or expressions in the text and add them to your vocabulary list!

unglücklich über	Aussehen, Erscheinung	überzeugt sein
lächeln, das Lächeln	aussehen, schauen	eindrucksvoll
sich (selbst-)sicher fühlen	sich entscheiden	einige Informationen über
bleichen	verfärbt	Urin
portugiesische Esel	Es funktioniert!	ich hoffe doch
heutzutage	stattdessen	Bleichset
verglichen mit	Bleichen in der Praxis	empfehlen
freiverkäufliche Produkte	eine große Gefahr	schädigen
Verbrennungen verursachen	die Haut irritieren	erfordern
vorsichtiger Umgang	Wirkstoff	Wasserstoffperoxid
die erzielte Bleichung	proportional zur Dauer	angewandt
Prozentsatz, Prozentanteil an	Prozedur, Vorgang	abhängig von
am häufigsten verwendet	vorbereiten	individuell angefertigte Tiefziehschiene
äußerst aggressiv	daher, deswegen	schützen

Zahnfleisch, Gingiva	sogenannt (-er/-en/-e)	Abdeckmembrane, Kofferdam
Nebenwirkungen	Nachteile	So weit ich weiß ...
die Oberflächen	empfindlich, sensibel für	Austrocknung, Dehydrierung
vermeiden	anwenden, auftragen	jede zweite oder dritte Nacht
im Falle von	ratsam, empfehlenswert	vor sich gehen, wirken
leicht eindringen in	oxidieren, bleichen	Flecken
Wie ist es mit ...?	im Gegensatz zu	verstärken
bewirken	harte Schleifmittel	mit der Zeit
zerstören	gelb erscheinen	empfindlich werden
Nahrungsmittel	Flüssigkeiten	schauen nach
wenig abrasiv/abschleifend	Kalziumperoxid	auf nicht-scheuernde Weise (etwas) entfernen

ACTIVITY 2

Have a look at the text and give an answer!

1. What does Mrs. Benning think about her teeth?
2. What is she convinced of?
3. Even the old Romans already tried to bleach teeth. How did they do it?
4. Why do dentists not recommend to buy "over the counter" products?
5. Why do the chemicals used for bleaching require careful handling?
6. What chemical is used for bleaching?
7. What is the active agent in carbamide peroxide?
8. What is the bleaching obtained dependent on?
9. How long does the procedure take?
10. How many methods of bleaching are applied in Dr. Zahn's surgery?
11. Explain the "home bleaching" method!
12. Explain the "waiting room" method!
13. The fastest method requires 10 minutes only. Explain the difference to the "waiting room" method!
14. Why must the procedure be repeated sometimes?
15. What is the cause of tooth sensitivity with some patients during and after treatment?
16. How can you avoid it?
17. How does bleaching work?
18. What do many people think about "whitening" toothpastes?
19. What can such toothpastes effectively do?

20. What do such toothpastes contain?
21. What is the danger of long-term use of such toothpastes?
22. What toothpastes should you look for?

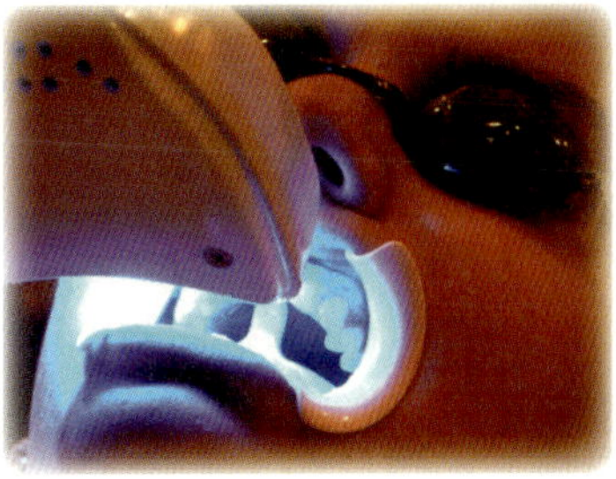

ACTIVITY 3

Please say it in English!

1. Viele Menschen sind nicht glücklich mit dem Aussehen ihrer Zähne.
2. Weiße Zähne sehen eindrucksvoll aus und man fühlt sich selbstsicherer, wenn man mit anderen Menschen spricht.
3. Statt (Instead of using) Eselurin, wie bei den alten Römern, verwendet man heutzutage Peroxide zum Bleichen der Zähne.
4. Freiverkäufliche Bleich-Sets sind natürlich billiger als das Bleichen in einer Zahnarztpraxis.
5. Produkte von Drogerien oder aus dem Internet sind nicht zu empfehlen.
6. Nur Zahnärzte oder Dental-Hygieniker machen es wirklich professionell.
7. Die Chemikalien, die zum Bleichen verwendet werden, sind aggressiv, irritieren die Haut und können Verbrennungen verursachen.
8. Die Gefahr ist groß, dass Zahnfleisch oder Zähne geschädigt werden, wenn man mit ihnen (den Produkten) nicht vorsichtig umgeht.
9. Zum Bleichen verwendet man Carbamidperoxid.
10. Der Wirkstoff in Carbamidperoxid ist Wasserstoffperoxid.
11. Der erzielte Bleicheffekt verhält sich proportional zur Anwendungsdauer und dem Anteil an verwendetem Wasserstoffperoxid.
12. Der Vorgang dauert entweder 8 Stunden, eine Stunde oder nur 10 Minuten.
13. Das hängt von der angewandten Methode ab.
14. Die am häufigsten angewandte Methode ist das „Home-Bleaching".
15. Der Zahnarzt fertigt für den Patienten eine individuelle Tiefziehschiene an.
16. Der Patient füllt diese Schiene vor dem Zu-Bett-Gehen mit einem 3%igen Wasserstoffperoxid-Gel und trägt sie über Nacht.
17. Die Behandlung ist nach ungefähr 7 bis 10 Nächten abgeschlossen.
18. Die schnellere Methode ist das Bleichen im Wartezimmer.
19. Es dauert nur eine Stunde.
20. Der Patient trägt die Schiene während er im Wartezimmer sitzt.
21. Die Schiene wird mit einem 12%igen Wasserstoffperoxid-Gel gefüllt.
22. Die schnellste Methode dauert nur 10 Minuten.

23. Allerdings muss man dann ein 35%iges Wasserstoffperoxid-Gel verwenden.
24. Weil diese Konzentration äußerst aggressiv ist, muss man das Zahnfleisch mit einem sogenannten „Kofferdam" schützen.
25. Diese Prozedur muss manchmal mehrmals wiederholt werden.
26. Der Zahnschmelz, das Dentin, die Pulpa und Oberflächen von Füllungen und künstlichen Zähnen werden nicht geschädigt.
27. Manchmal reagieren Zähne nach der Behandlung empfindlich auf heiß oder kalt.
28. Das kommt von der Dehydrierung der Zähne.
29. In solchen Fällen ist es ratsam, das Gel nur jede zweite oder dritte Nacht aufzutragen.
30. Im Falle eines Bleichens in der Praxis sollte man 2 bis 3 Tage warten, bevor man wieder ein Gel aufträgt.
31. Der Sauerstoff im Wasserstoffperoxid kann leicht in den Zahnschmelz eindringen.
32. Innerhalb des Zahnschmelzes werden Flecken und Verfärbungen „oxidiert", das heißt, gebleicht.
33. „Weißmachende" Zahncremes verstärken nicht das Weiß der Zähne.
34. Jedoch entfernen sie sehr effektiv Flecken, die durch Kaffee, Tee und Tabak verursacht sind.
35. Mit der Zeit zerstören die Schleifpartikel in diesen Zahncremes den Zahnschmelz und die Zähne erscheinen gelb.
36. Suchen Sie nach Zahnpasten, die minimalabrasiv sind und Fluorid sowie Kalziumperoxid und Xylitol enthalten.

Patient: "How much is it to have this tooth pulled?"

Dentist: "200 Euros."

Patient: "200 Euros for just a few minutes work???"

Dentist: "I can extract it very slowly, if you like. Then it will take longer."

(How much is it? = Wieviel kostet es?; to have pulled = ziehen lassen; just = lediglich, nur; a few = ein paar; it takes longer = es dauert länger)

>>> ACTIVITY 4 >>>

Work with a partner. Ask each other these questions!

1. Are you happy about the appearance of your teeth?
2. Have you ever assisted in cosmetic dental treatment like bleaching?

3. What's the difference between professional dental cleaning and cosmetic dental treatment?
4. What's the most efficient way to get white teeth?
5. How is bleaching done?
6. Are there disadvantages or side-effects in bleaching?
7. Why do dentists not recommend buying "over-the-counter" bleaching sets?

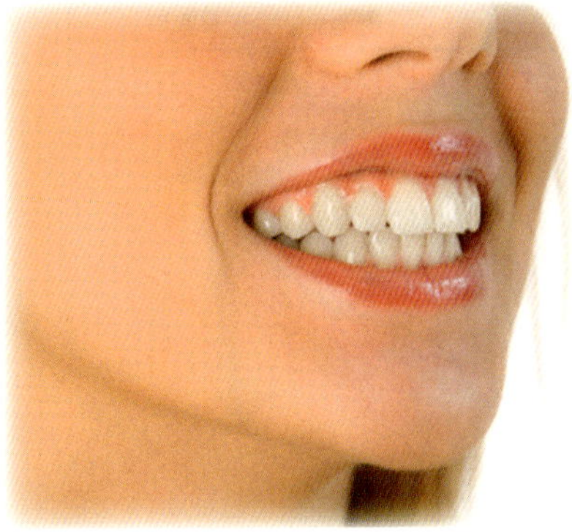

ACTIVITY 5

Make this conversation with a partner!

Assistent	Patient
1. Begrüßen Sie die Patientin und sagen Sie ihr, dass Sie sie über kosmetische Zahnbehandlung informieren wollen.	
	2. Fragen Sie nach dem Unterschied zur professionellen Zahnreinigung.
3. Erklären Sie den Unterschied.	
	4. Fragen Sie, ob Sie auch ein Bleich-Set für Zuhause versuchen können.
5. Raten Sie der Patientin davon ab und begründen Sie dies ausführlich.	
	6. Sie möchten gerne wissen, welche Methoden zur Zahnaufhellung es gibt und wie viel Zeit die jeweilige Methode in Anspruch nimmt.
7. Nennen und erklären Sie die Methoden, die in Ihrer Praxis angewandt werden.	
	8. Fragen Sie nach eventuellen Nebenwirkungen oder Nachteilen des Bleichens.
9. Geben Sie eine Antwort.	
	10. Bedanken Sie sich für die Informationen.
11. Erwidern Sie den Dank.	

EXTENDED VOCABULARY LIST	
aesthetic	ästhetisch
aesthetics	Ästhetik
alternative	Alternative
appear	erscheinen, zum Vorschein kommen, auftauchen
appearance	das Äußere, das Aussehen, das Erscheinungsbild
appearances **Appearances can deceive**	der Schein, der Anschein Der Schein kann trügen.
aware **I'm aware of that!**	bewusst Ich bin mir dessen bewusst!
awareness	Bewusstsein
compare (to or with)	vergleichen mit, prüfen, steigern (Grammatik)
comparison	Vergleich, Gegenüberstellung
confidendence	Vertrauen, Zuversicht, Selbstsicherheit, Geheimnis
confident	sicher, überzeugt, zuversichtlich, selbstsicher
convince	überzeugen, überreden
convinced (of)	überzeugt (von)
cosmetic	kosmetisch
cosmetics	Kosmetik
couple **a couple of**	ein Paar, Ehepaar, verkuppeln, verheiraten einige, ein paar
cosmetic dentistry	kosmetische Zahnheilkunde
dentures	Prothesen, künstliches Gebiss, „dritte Zähne“
differ (from) **I beg to differ!** **I differ about it.**	sich unterscheiden (von), abweichen (von) Da bin ich anderer Meinung! Ich denke anders darüber.
difference	Unterschied
efficacy	Wirksamkeit, Wirkungskraft

EXTENDED VOCABULARY LIST	
efficiency	Leistungsfähigkeit, Wirksamkeit
efficient	wirksam, effizient, rationell
either ... or	entweder ... oder
fact	Tatsache, Fakt
false	falsch, unrichtig, treulos, unwahr
fluoride	Fluorid
fluorine	Fluor
happy (about)	glücklich, zufrieden mit
harm	Schaden, Verletzung, Schaden anrichten, verletzen
harmful	schädlich, nachteilig
harmless	ungefährlich, unschädlich, harmlos
hydrogen **hydrogen peroxide** **hydrogen sulphide**	Wasserstoff Wasserstoffperoxid Schwefelwasserstoff
impress	beeindrucken
impressing	beeindruckend
impression	Eindruck, Abdruck, Ahnung, unbestimmtes Gefühl
impressive	eindrucksvoll
luck **lucky** **good luck** **bad luck** **I should be so lucky!**	Glück Glück haben Glück Pech Das möchte ich mal erleben!
main **the main thing** **in the main**	Haupt-, Hauptleitung (Strom) die Hauptsache im Großen und Ganzen
offer	anbieten
office **in office**	Büro, Amt, in USA auch Praxis eines Zahnarztes hier: in der Praxis
pharmacy (US), chemist's shop (UK)	Apotheke
quiet	still, leise, ruhig

EXTENDED VOCABULARY LIST	
quit	aufhören mit, kündigen, beenden
Quit smoking!	Hör' mit dem Rauchen auf!
I've given him a notice to quit.	Ich habe ihm gekündigt.
Let's call it quits.	Jetzt sind wir quitt.
remove	entfernen
quite	ziemlich
think	denken
What do you think of it?	Was halten Sie davon?
I thought about it.	Ich dachte daran
I think it over.	Ich werde es überdenken/darüber nachdenken
Don't think me rude.	Halten Sie mich nicht für unhöflich.
That's what you think!	Das denkst Du wohl!
Who would have thought it!	Wer hätte das gedacht!
Think nothing of it!	Das ist nicht der Rede wert!
to make something look like	etwas aussehen lassen wie
whereas	während, wohingegen

8 TOOTH DECAY

Barbara is a qualified dental employee. Her English teacher asked her to give a lecture in English. The topic of her lecture is: "How caries starts and how to stop it". This is what she wrote ...

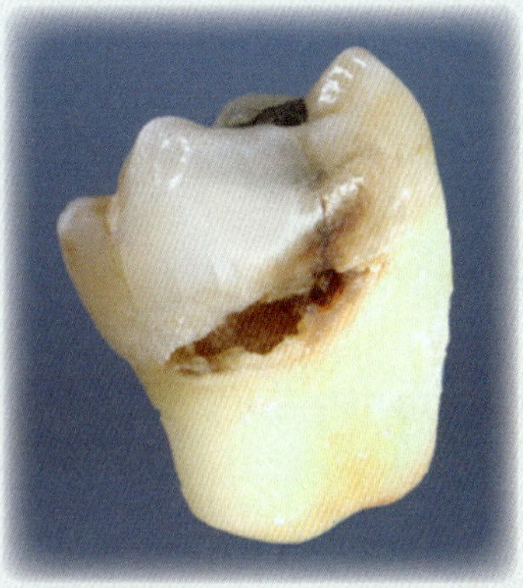
Tooth Decay

"There are many people who think that dental problems are mainly the result of negligent dental care. However, negligent dental care is only one of several reasons for tooth decay. Quite a few people are having problems although they clean their teeth twice a day, use all kinds of dental care products like toothpastes containing fluoride, dental floss, sugar-free dental gum and mouthwashes. In that case it might be a matter of genetics. Another reason is food. If you never eat sweet things like cakes, chocolate bars, ice cream, sweetened soft drinks and lots more, tooth decay will only have little chance. However, most of the food you eat contains sugar. You just cannot avoid it. Sugar is the worst enemy of healthy teeth. Why is that so?

Everyone has got bacteria in the mouth. That's normal. Most of the bacteria are not harmful at all. However, if you eat sweet things, the bacteria start turning the sugar into a very aggressive acid. The acid then starts attacking the gums and destroys the enamel on the outside of the teeth. The dental enamel is the most efficient protection for the teeth. If the dental enamel becomes destroyed, the acid makes the teeth decay quickly. The result is that cavities are formed.

How can you stop or even better, prevent it?

There are several ways how to stop this process: One way is, to kill the bacteria in your mouth by using a mouthwash. If there are no bacteria left, no harmful acid will arise. The other way is, to use dental gum or cheese after every meal. Dental gum does not kill the bacteria but reduces the acidity. Then, regular tooth cleaning and flossing every day is important, too. All these things do not cost much but make a big difference to your teeth!"

ACTIVITY 1

Find these words or expressions in the text and add them to your vocabulary list!

sie werden gebeten	ein Referat halten	das Thema
hauptsächlich	das Ergebnis, die Folge	nachlässige Zahnpflege
einer von verschiedenen Gründen	Karies, Zahnverfall, Zahnfäule	ziemlich viele Menschen
obwohl	alle möglichen	Zahnpflegeprodukte
Fluorid	Zahnseide	zuckerfrei
Zahnkaugummi	Mundspülungen	in diesem Fall
es könnte sein	genetisch bedingt	die Ernährung
Schokoriegel	gezuckerte, alkoholfreie Getränke	eine geringe Chance
Nahrungsmittel	man kann es nicht vermeiden	der schlimmste Feind
sie sind überhaupt nicht schädlich	den Zucker umwandeln in	aggressive Säure
das Zahnfleisch angreifen	den Zahnschmelz zerstören	auf der Außenseite
ein wirksamer Schutz	zerstört werden	langsam zersetzen/ verfaulen
es bilden sich Löcher	etwas verhindern	den Vorgang aufhalten
die Bakterien abtöten	hier: übrig	schädliche Säure
hier: entstehen	nach jeder Mahlzeit	den Säuregehalt reduzieren
dann ...	regelmäßiges Zähneputzen	es macht viel aus für ...

ACTIVITY 2

Look at the text again and give an answer!

1. What is the topic of the lecture?
2. Is having dental problems always the result of negligent dental hygiene?
3. What food should you avoid?
4. Sugar is the enemy of healthy teeth. How come?
5. How can you reduce harmful acid in your mouth?
6. What effect does dental gum have?
7. What can you do to keep your teeth healthy?

ACTIVITY 3

Translate these sentences!

1. Wie fängt Karies an und wie kann man sie aufhalten?
2. Zahnprobleme sind nicht nur die Folge nachlässiger Zahnpflege.
3. Für Karies gibt es verschiedene Gründe.
4. Nachlässige Zahnpflege ist nur einer von mehreren Gründen.
5. In manchen Fällen sind Zahnprobleme genetisch bedingt.
6. Ich verwende alle möglichen Zahnpflegeprodukte.
7. Sie sollten nur Zahncreme verwenden, die Fluorid enthält.
8. Fluorid härtet und schützt den Zahnschmelz.
9. Verwenden Sie auch Zahnseide und zuckerfreien Zahnkaugummi!
10. Zucker ist der schlimmste Feind der Zähne.
11. Süße Nahrungsmittel sind sehr schädlich für die Zähne.
12. Ohne Zucker hat Karies nur geringe Chancen.
13. Normalerweise sind die meisten Mundbakterien harmlos.
14. Jedoch verwandeln Mundbakterien Zucker in eine sehr aggressive Säure.
15. Die Säure greift das Zahnfleisch an und zerstört den Zahnschmelz.
16. Der Zahnschmelz ist der wirksamste Schutz für die Zähne.
17. Wenn der Zahnschmelz zerstört wird, verursacht die Säure Zahnverfall.
18. Die Folge (davon) ist, dass sich Löcher bilden.
19. Wie kann man dies aufhalten – oder besser – verhindern?
20. Es gibt verschiedene Wege/Möglichkeiten, diesen Prozess aufzuhalten.
21. Die Bakterien im Mund lassen sich durch Mundspülungen abtöten.
22. Eine Möglichkeit ist, nach jeder Mahlzeit einen Zahnkaugummi zu verwenden.
23. Zahnkaugummi reduziert den Säuregehalt.
24. Regelmäßige Zahnreinigung und Verwendung von Zahnseide sind ebenso wichtig.

Patient: “What would an X-ray of my head show?”

Doctor: “Absolutely nothing.”

(X-ray = Röntgenaufnahme; show = zeigen, ergeben; absolutely nothing = rein gar nichts)

ACTIVITY 4

Make this conversation with a partner!

Student A	Student B
1. Fragen Sie nach dem Thema des Referates.	
	2. Das Thema lautet: „Wie Karies entsteht und wie man sie aufhalten kann.“
3. Behaupten Sie, dass Zahnprobleme allein wegen nachlässiger Zahnpflege entstehen.	
	4. Entgegnen Sie, dass manche Menschen trotz guter Zahnpflege immer Zahnprobleme haben und nachlässige Zahnpflege nur einer von mehreren Gründen für Zahnprobleme ist.
5. Fragen Sie, was ein weiterer Grund sein könnte.	
	6. Geben Sie darauf eine Antwort. Weisen Sie auch darauf hin, dass falsche Ernährung, die viel Zucker enthält, ein weiterer Grund ist.
7. Fragen Sie, weshalb dies so ist?	
	8. Erklären Sie den Zusammenhang zwischen Zucker in der Nahrung und der Entstehung von Karies.
9. Fragen Sie, wie man den Zahnverfallsprozess aufhalten oder besser noch, vermeiden könnte.	
	10. Geben Sie eine Antwort auf diese Frage.

EXTENDED VOCABULARY LIST	
lecture **to give a lecture on**	Referat, Vortrag, Vorlesung ein Referat halten über
arise (from/out of)	entstehen, ergeben (Probleme) aus/von
harden	härten, stählen, abhärten, festigen, stärken
neglect	vernachlässigen, versäumen, nicht befolgen (advice)
negligent **Don't be negligent of it!**	nachlässig, unachtsam, fahrlässig Vernachlässige es nicht!
bar	Stange, Tafel (Schokolade), Riegel, Sperre, Hindernis
little **in a little while**	wenig, klein, gering, kurz, kaum in kurzer Zeit
avoid	vermeiden, meiden, aus dem Wege gehen
avoidable	vermeidbar
bad/worse/worst	schlimm, schlimmer, am schlimmsten
turn	drehen, Drehung, Kurve; hier: verwandeln
decay	zersetzen, verfaulen, schlecht werden, verblühen
tooth decay	Karies

A seventy year old patient went for a regular check-up of his teeth.

After the examination the dentist smiled at him and said: "Your teeth are good for the next 50 years!"

The patient replied: "I won't live any longer than another 20 years! What will my teeth do without me when I'm dead?"

(to smile at s.o. = jemanden anlächeln; to be good for = sie halten noch weitere ...; reply = erwidern; to have to die = sterben müssen)

9 TROUBLE WITH THE GUMS

Colonel Frank N. Furter is a NATO officer, presently doing his service at a NATO air base in Germany. The Army Health Clinic referred him to Dr. Zahn's dental office as Dr. Zahn has got a contract with the US-Army. Colonel Furter urgently needs a dental appointment. Dr. Zahn's assistant gave him an appointment for the following day. Mr. Furter is entering the surgery ...

Mr. Furter Good morning. I was told you speak English in your office. My name is Furter. Frank N. Furter. I talked to you on the phone yesterday.

Assistant Ah, yes, sir. You've got an appointment for 10.30, right?

Mr. Furter Yes, that's right.

Assistant Could you please fill in this form here? We need a few personal details from you. You have got dental insurance with the Army?

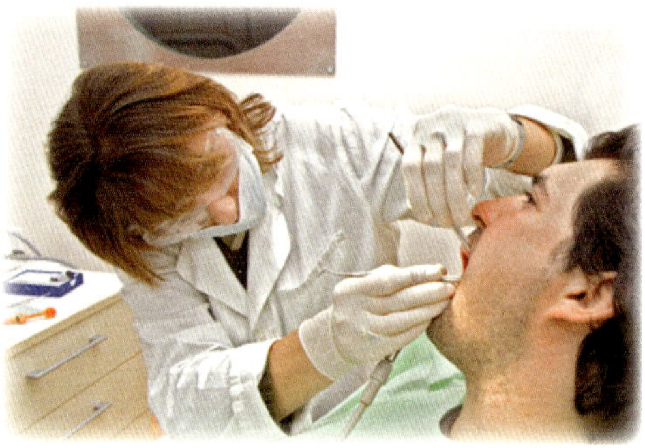

Mr. Furter Sure. (He completes the form and hands it back to the assistant ...)

Assistant Oh, by the way, Mr. Furter. What does the middle initial "N." in your name stand for?

Mr. Furter It stands for "Norman". That's my middle name.

Assistant I see. OK. Could you please take a seat in our waiting room?

Mr. Furter Where is it?

Assistant Just round the corner. It's the first door on the left, opposite the restrooms. I'm afraid you will have to wait a bit. We are quite busy today. But, don't worry; it won't take longer than 20 minutes.

I'll call you as soon as the doctor can see you.

Mr. Furter Right.

After Dr. Zahn has seen Colonel Furter, the doctor sends him to another treatment room. There, one of Dr. Zahn's assistants takes care of Mr. Furter ...

Assistant Hello.

Mr. Furter Hi.

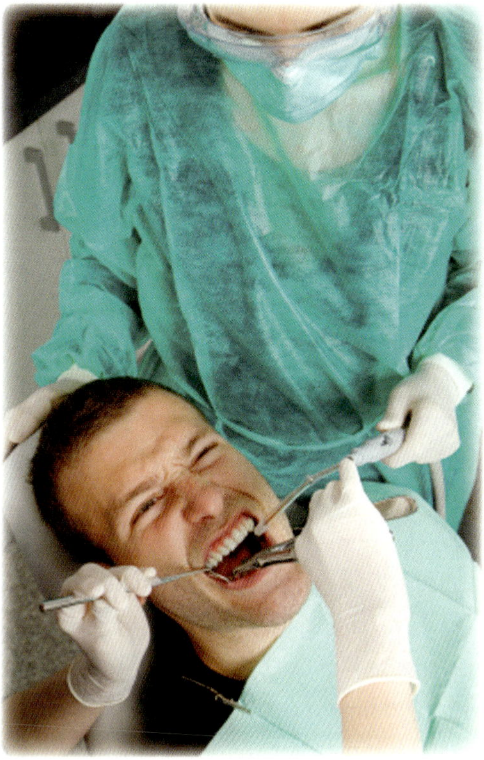

Assistant The doctor already told me about your problem. It's your gums, right?

Mr. Furter Yes, I'm having lot of trouble with my gums.

Assistant Could you open up nice and wide and turn your head towards me?

Well, there's a lot of red inflammation around your gums. Do your gums bleed when you brush your teeth?

Mr. Furter Yes, they do. Tooth brushing is quite painful.

Assistant Your gums are very red and inflamed. You seem to have a very bad inflammation of your gums. We call it gingivitis. But, when all the tartar is removed, the bleeding should stop. I think, regular flossing would be a good idea, too.

Mr. Furter Flossing?

Assistant Yes, flossing helps to prevent gingivitis. You should floss between all your teeth, especially where your teeth meet your gums. There isn't anything more efficient to remove tartar and plaque than flossing, you know.

Mr. Furter My teeth are also very sensitive to heat and cold.

Assistant That's because your gums have receded around your teeth. Your tooth necks are exposed, so your dental nerves are unprotected. That's why your teeth are so sensitive to pain.

After the treatment

Assistant Ok, Mr. Furter. That's it. You should make another appointment in about one week. Then we'll see what your gums are like ...

UNIT 9

ACTIVITY 1

Find these words or expressions in the text and add them to your vocabulary list!

Schwierigkeiten, Probleme	zurzeit, momentan	seinen Dienst ableisten
Abkürzung für ***N**orth **A**tlantic **T**reaty **O**rganisation*	Luftwaffenstützpunkt	Ärztliches/Medizinisches Versorgungszentrum (US)
überweisen/weiterleiten	Zahnarztpraxis (US)	einen Vertrag haben
dringend	ein Zahnarzttermin	mir wurde gesagt ...
am Telefon sprechen	stimmt´s?	ein Formular ausfüllen
persönliche Angaben	Zahnversicherung	Sicher!, Klar!
der Anfangsbuchstabe in der Mitte	Wofür steht er?	zweiter Vorname (US)
gleich um die Ecke	Erste Türe links	gegenüber
unsere Praxis ist ziemlich voll	Keine Sorge!	sobald als ...
nach kurzer Zeit	sich kümmern um ...	Hallo. (US)
in meine Richtung	entzündete Stellen	ziemlich schmerzhaft
rot und entzündet	eine schlimme Entzündung	Zahnfleischentzündung
regelmäßiges Verwenden von Zahnseide	Gingivitis, Zahnfleisch-erkrankung	verhindern
Zahnseide benutzen	zwischen den Zähnen	insbesondere
wo Zahn und Zahnfleisch aufeinandertreffen	Es gibt nichts Wirksameres als ...	empfindlich sein auf
Hitze und Kälte/heiß und kalt	das Zahnfleisch ist zurück-gewichen	die Zahnnerven
schutzlos	deshalb, deswegen, darum	schmerzempfindlich
Das war's!	Dann sehen wir einmal ...	wie das Zahnfleisch aussieht

Dentist to his assistant: “My patient, Miss Miller, is the only woman I ever loved.”
Assistant: “Why don’t you ask her to marry you?”
Dentist: “I can’t afford to marry her.”
Assistant: “Why?”
Dentist: “She’s my best patient!”

(ever = jemals; marry = heiraten; afford = leisten)

ACTIVITY 2

Look at the text again and give an answer!

1. Who referred Colonel Furter to Dr. Zahn?
2. What did he urgently need?
3. The assistant asked Colonel Furter to fill in something. What should he do?
4. Has he got dental insurance?
5. What does the middle initial "N" in his name stand for?
6. Where is the waiting room located?
7. Colonel Furter will have to wait a while. Why?
8. What is causing him trouble?
9. The assistant asked him to open up his mouth. What did she see?
10. What happens when he brushes his teeth?
11. When should the bleeding stop?
12. What did the assistant advise him to do regularly?
13. Why is flossing so important?
14. Colonel Furter's teeth are sensitive to heat and cold. Why?
15. Why should Colonel Furter make another appointment?

Professional dental cleaning

Dentist to patient: "No, we give no discounts for gaps between your teeth! ".

Cosmetic dental treatment

Dentist: "After bleaching your teeth, you'll be another man! "

Patient: "Fine, but don't forget to send the bill to the other man."

(discount = Nachlass; gaps between teeth = Zahnlücken; another = ein weiterer, ein anderer; bill = Rechnung)

ACTIVITY 3

Please, translate these sentences!

1. Wer hat Sie zu unserer Zahnarztpraxis überwiesen?
2. Ich glaube, ich habe gestern am Telefon mit Ihnen gesprochen.
3. Könnten Sie bitte dieses Formular ausfüllen?
4. Ich benötige ein paar persönliche Angaben von Ihnen.
5. George W. Bush war der Name eines amerikanischen Präsidenten.
6. Das „W“ in seinem Namen steht für „Walker“.

7. Es ist sein sog. Mittelname (eine angelsächsische Besonderheit)
8. Unser Wartezimmer ist hier, gleich um die Ecke!
9. Es ist die erste Türe links.
10. Die Praxis ist heute ziemlich voll.
11. Es wird heute etwas länger dauern.
12. Bitte gehen Sie in den Behandlungsraum 2.
13. Eine Helferin wird sich dort um Sie kümmern.
14. Blutet Ihr Zahnfleisch beim Putzen?
15. Ihr Zahnfleisch ist sehr entzündet.
16. Eine Zahnfleischentzündung wird auch Gingivitis genannt.
17. Das Bluten müsste aufhören, wenn der Zahnstein entfernt ist.
18. Regelmäßiges Verwenden von Zahnseide wäre eine gute Idee.
19. Mit Zahnseide kann man eine Zahnfleischerkrankung verhindern.
20. Es gibt nichts Wirksameres um Zahnstein und Zahnbelag zu entfernen, als das Verwenden von Zahnseide.
21. Besonders (dort), wo Zähne und Zahnfleisch aufeinander treffen.
22. Das Zahnfleisch um Ihre Zähne herum, ist bereits zurückgewichen.
23. Die Nerven Ihrer Zähne sind ungeschützt.
24. Deshalb sind Ihre Zähne so schmerzempfindlich.
25. In einer Woche sehen wir, wie es um Ihr Zahnfleisch steht.

A man, frustrated from women, says to his friend: "I'd like to be a dentist!"

Friend: "Why?"

Man: "Because it's the only profession where you can tell a woman when to open her mouth and when to shut it ..."

(profession = akademischer Beruf; shut = schließen, zumachen, hier auch: den Mund „halten")

ACTIVITY 4

Ask your partner a few questions!

1. Do you speak English in your dental surgery/office?
2. Which questions would you ask a foreign patient who visits your surgery/office for the first time?
3. You are sitting at the reception. A patient would like to know where the waiting room is. How would you tell him where it is?
4. Americans very often use a middle initial in their names. What does it stand for?
5. Have your gums ever given you any trouble?

6. How would you diagnose gingivitis?
7. What is gingivitis?
8. Do you know an efficient way to prevent tartar and plaque?
9. A patient's gums have receded around the teeth. What consequences can it have?

ACTIVITY 5

Make a conversation with a partner!

Assistent	Patient
	1. Sie sind eine Englisch sprechende, neue Patientin. Sie betreten die Praxis und grüßen.
2. Begrüßen Sie sie auf Deutsch und fragen Sie, was Sie für die Patientin tun können.	
	3. Ihnen wurde gesagt, dass in dieser Praxis auch Englisch gesprochen wird. Stellen Sie sich vor und sagen, dass Sie einen Termin um ... Uhr haben.
4. Bestätigen Sie den Termin. Sie benötigen einige persönliche Angaben. Bitten Sie die Patientin, ein Formular auszufüllen.	
	5. Sie kommen der Bitte nach.
6. Fragen Sie, ob die Patientin krankenversichert ist oder bar bezahlt.	
	7. Sie bezahlen bar.
8. Sie bitten die Patientin, die Honorarvereinbarung zu unterschreiben und dann im Wartezimmer Platz zu nehmen.	
	9. Sie fragen, wo es sich befindet.
10. Sie erklären, wo es ist und dass es etwas dauern wird, weil heute viel los ist. Sie sagen der Patientin, wie lange sie voraussichtlich warten muss.	
	11. Sie erwidern, dass es kein Problem ist, etwas warten zu müssen.
12. Sie rufen die Patientin auf, und bitten Sie in das Behandlungszimmer 2 zu gehen. Dort wartet eine ZFA auf sie. Sagen Sie ihr, dass Doktor Zahn gleich kommen wird.	
	13. Sie betreten das Behandlungszimmer und grüßen die dort wartende ZFA.

14. Sie sind die dort wartende ZFA. Sie grüßen die Patientin und bitten Sie, auf dem Behandlungsstuhl Platz zu nehmen und ihren Mund weit aufzumachen. Falls die Patientin schlucken muss (= need to swallow), soll sie es sagen. Sie stellen fest, dass das Zahnfleisch der Patientin stark entzündet ist. Fragen Sie, ob es beim Zähneputzen blutet.

15. Bejahen Sie dies.

16. Sie stellen fest, dass sich um die Zähne herum viel Zahnstein befindet. Sagen Sie der Patientin, dass nach der Entfernung des Zahnsteins das Zahnfleischbluten aufhören sollte.

17. Fragen Sie, wie Sie der Entstehung von neuem Zahnstein vorbeugen können.

18. Weisen Sie die Patientin darauf hin, wie wichtig die Verwendung von Zahnseide ist und dass es nichts Besseres gibt, um Zahnstein und Zahnbelag vorzubeugen.

19. Ihre Zähne reagieren empfindlich auf Kaltes und Warmes. Sagen Sie dies der ZFA.

20. Erklären Sie der Patientin, weshalb das so ist.

21. Erwidern Sie etwas darauf.

22. Sagen Sie der Patientin, dass sie sich in etwa einer Woche einen weiteren Termin geben lassen soll. Bis dahin wird man sehen, ob, hinsichtlich ihrer Zahnfleischentzündung, eine Besserung eingetreten ist.

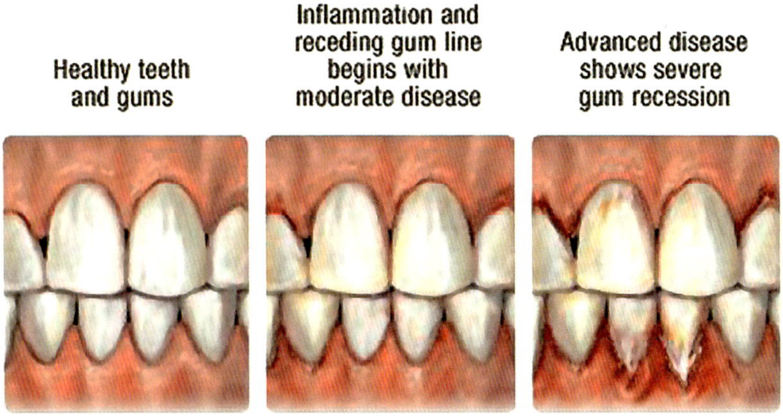

EXTENDED VOCABULARY LIST	
advanced disease	fortgeschrittene Erkrankung
by the way	übrigens
consequences	Folgen, Konsequenzen
contract	Vertrag, Abkommen, Kontrakt
details	Einzelheiten, Details, Angaben
exposed	freiliegend, freigelegt
Health Clinic	Ärztliches Versorgungszentrum, Gesundheitszentrum (US)
initial **initial letter** **middle initial**	anfänglich, Anfangs-, Initiale Anfangsbuchstabe Anfangsbuchstabe des Mittelnamens George W. Bush
insurance company **medical insurance** **dental insurance** **privat insurance**	Krankenversicherungs(gesellschaft) Krankenversicherung Zahnversicherung Privatversicherung
insure (against)	versichern (gegen)
insured	versichert
insured person	der/die Versicherte (Person)
insurer	Versicherer
like **What is he like?** **What is it like?** **That´s not like me.** **Something like that.** **It looks like rain.**	ähnlich, gleich Was ist er für ein Mensch? Wie sieht es aus?, Wie ist es? Das ist nicht meine Art. So etwas Ähnliches. Es sieht nach Regen aus.
meet	hier: treffen auf, zusammentreffen
moderate disease **severe disease**	leichte, mittelgradige, mittelschwere Erkrankung schwere Erkrankung
partially obscuring	teilweise verdeckend
receding gum line	Zahnfleischrückgang, zurückweichende Zahnfleischlinie
restroom (US) **lavatory (UK)** **toilet (UK)**	Toilette

EXTENDED VOCABULARY LIST	
sensible	vernünftig (aufpassen, nicht verwechseln mit „sensibel“!!!)
sensitive	sensibel, empfindlich
sir/madam	Herr/Dame
take	hier. brauchen, benötigen, erfordern
tooth neck	Zahnhals
unprotected	ungeschützt
with	hier: bei (der Armee)

Young girl to father: “Daddy, when I grow up shall I become a heart-doctor or a tooth-doctor?”

Father: “You should become a dentist”

Girl: “Why, father?”

Father: “We have only one heart, but 32 teeth! “

(to grow up = erwachsen werden; shall I? = soll ich?; become = werden)

10 DENTAL FEAR

Mr. Hugh Carter is an Australian computer specialist working in Germany. He has had trouble with his wisdom teeth for some time now. They must be pulled. He already called Dr. Zahn's dental surgery and got an appointment for Monday, October 15th, at 13.30 hours. However, there is another problem: He also suffers from a severe phobia about dental appointments ...

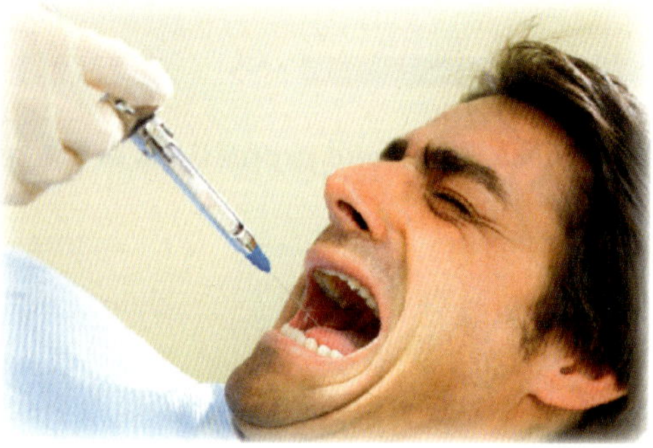

Assistant Hello, Mr. Carter. How are you today?

Mr. Carter Not great at all.

Assistant Oh, don't worry. Dr. Zahn is very experienced.

Mr. Carter That's nice of you to say, but thinking of my dental appointment has stopped me from sleeping for almost a week now. Ever since I was a child I have been suffering from dental fear, you know.

Assistant Dental fear is a normal condition. Many people suffer from that.

Mr. Carter Do they also have a feeling of sickness, diarrhoea, difficulty breathing and muscle tension?

Assistant Well, some people have that, too.

Mr. Carter If I only imagine entering the examination room with that typical smell, or imagine the sound of the drill, I just start sweating. I hate having any pain!

Assistant Well, you will not have any pain, Mr. Carter. We know this problem. There are different ways to make dental treatment possible for you.

Mr. Carter Which are they? Local anaesthesia?

Assistant In your case local anaesthesia won't do. Dr. Zahn can put you in half sleep, for example. He often does that if the treatment does not exceed 30 minutes. There also is the possibility of dental treatment under general anaesthesia.

Mr. Carter What about hypnosis? I've heard some dentists do that, too. I'm hypnotizable, you know, and can cooperate with Dr. Zahn in spite of my fear.

Assistant We are not able to offer you that. In Germany, doctors need a particular qualification to do that.

Mr. Carter Well, I guess, half sleep would be a good idea, too ...

ACTIVITY 1

Find these words and expressions in the text and add them to your vocabulary list!

Angst vor dem Zahnarzt	Probleme haben mit	Weisheitszähne
seit einiger Zeit	ziehen/herausziehen	leiden an
schwer, schlimm	Zahnbehandlungsphobie	Wie geht´s Ihnen heute?
Gar nicht gut!, Nicht so toll!	sehr erfahren sein	es hält mich vom Schlafen ab
seit, seither	leiden an	Angst vor dem Zahnarzt
ein normaler Zustand	ein Gefühl des Ekels/ Brechreizes/Widerwillens	Durchfall, Diarrhoe
Atemprobleme	Muskelverspannung	Wenn ich mir nur vorstelle ...
ein typischer Geruch	das Geräusch eines Bohrers	anfangen zu schwitzen
Schmerz	Örtliche Betäubung, Lokalanästhesie	in Ihrem Falle
es wird nicht reichen	in einen Dämmerschlaf versetzen	übersteigen, übertreffen
Behandlung in Vollnarkose	Wie wäre es mit ...?	Hypnose
Man kann mich hypnotisieren	zusammenarbeiten mit, mitwirken bei, kooperieren mit	trotz meiner Angst
Wir sind nicht in der Lage	etwas anbieten	besonder/e/es/r
Qualifikation, Voraussetzung	ich schätze/nehme an	eine gute Idee/Vorschlag

ACTIVITY 2

Read the text again and give an answer!

1. Who is Mr. Carter?
2. What dental problem does he have?
3. He has another big problem. What is it?
4. For how long has he been suffering from it?
5. Describe the symptoms of his problem!
6. Fear of pain is the reason for his phobia. Can Dr. Zahn help him?
7. What about hypnosis?

ACTIVITY 3

Say it in English!

1. Ihre Weisheitszähne müssen gezogen werden.
2. Ich leide an einer Zahnbehandlungsphobie.
3. Ich fühle mich heute nicht so toll!
4. Machen Sie sich keine allzu großen Sorgen!
5. Wir kennen uns mit so etwas aus./Wir sind sehr erfahren.
6. Meine Angst hält mich seit Wochen vom Schlafen ab.
7. Viele Menschen leiden an Angst vor dem Zahnarzt.
8. Viele Menschen bekommen Durchfall, Atemprobleme, Muskelverspannungen oder Brechreiz, wenn sie nur daran denken, einen Behandlungsraum mit diesem typischen Geruch betreten zu müssen.
9. Manche Patienten beginnen zu schwitzen, wenn sie das Geräusch des Bohrers hören.
10. Viele Patienten wollen keine Schmerzen verspüren.
11. Es gibt verschiedene Therapien, um solchen Patienten eine zahnmedizinische Behandlung zu ermöglichen.
12. Zum Beispiel gibt es die Möglichkeit einer örtlichen Betäubung oder eines Dämmerschlafs.
13. Es gibt auch die Möglichkeit einer Vollnarkose oder Hypnose.
14. Nur wenige Ärzte bieten eine Hypnose an.
15. Ärzte benötigen eine bestimmte Qualifikation dafür.

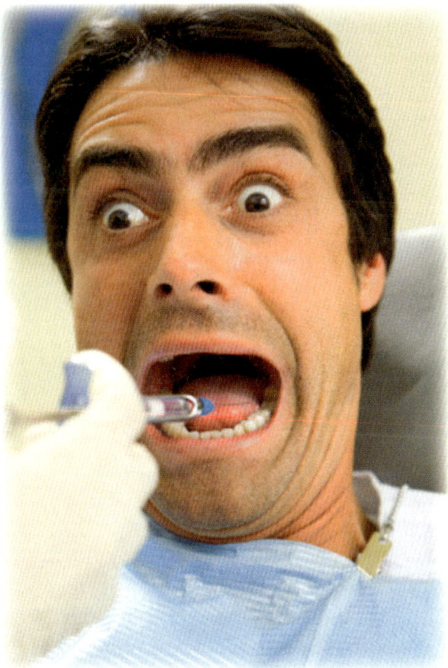

UNIT 10

ACTIVITY 4

Work with a partner. Ask her or him questions!

1. Do you have any experience with patients suffering from dental fear?
2. How do you treat such patients in your dental surgery/dental office?
3. Which symptoms may some patients suffer from who have this phobia?
4. What about you? Do you have this problem, too?
5. Do some of your patients feel sick if the dentist has to use a drill?
6. What about local anaesthesia? Do you have a problem with injections in your mouth?
7. What kinds of anaesthesia do you practise in your dental surgery?

ACTIVITY 5

Make a conversation with a partner!

Assistent	Patient
1. Sie bitten die Patientin, auf dem Behandlungsstuhl Platz zu nehmen, und fragen sie, wie es ihr geht.	
	2. Es geht Ihnen gar nicht gut und sagen auch warum. Sie haben eine panische Angst vor dem Zahnarzt und vor Schmerzen.
3. Sie sagen der Patientin, dass sie mit diesem Problem nicht alleine ist. Sie haben viel Erfahrung mit diesem Problem.	
	4. Sie schildern Ihre Symptome.
5. Sie kennen das.	
	6. Allein bei dem Gedanken an den Geruch in einer Zahnarztpraxis oder dem Geräusch eines Bohrers wird Ihnen schlecht.
7. Beruhigen Sie die Patientin und sagen Sie ihr, dass sie keinerlei Schmerzen verspüren wird. Weisen Sie sie auf die verschiedenen Betäubungsmöglichkeiten hin.	

EXTENDED VOCABULARY LIST	
anaesthesia	Narkose, Betäubung
difficulty breathing	Atemprobleme, Schwierigkeiten beim Atmen
ever since **hardly ever** **ever and again**	seitdem fast nie immer wieder
general anaesthesia	Vollnarkose
guess **I guess so!** **I guess ...**	schätzen, erraten, vermuten, ahnen, annehmen, glauben Vermutlich! Ich denke/glaube ... (US)
half sleep	Dämmerschlaf
in spite of	trotz, obwohl
local anaesthesia	Lokale Betäubung, Lokalanästhesie, Teilnarkose
local anaesthetic	Lokalanästhetikum, Mittel zur örtlichen Betäubung
particular **to be particular**	besondere/r/s/bestimmte (z. B. Menschen) wählerisch sein
sick **I feel sick.** **a feeling of sickness**	krank, schlecht Mir ist schlecht/übel Übelkeit, ein Brechreiz
smell **to smell of ...**	(eher schlechter) Geruch; (wohlriechend – fragrance) riechen, duften, stinken nach
stop from	abhalten von
tension	Spannung, Anspannung, Verspannung, gespannte Lage

11 FOOD AND TEETH

Dr. Zahn would like to make a flyer for his English speaking patients. The headline should read: "Foods That Destroy Your Smile". As all of his qualified dental employees are able to speak English, Dr. Zahn asks them to write the text of the flyer. This is what his assistants wrote:

Foods That Destroy Your Smile

"Sugar is the worst enemy of your teeth! If you eat food, only consisting of fat and protein, you may get plaque, but never get any real problems like caries. Fat and protein do not attack the dental enamel. The bacteria in your mouth do not use the fat and protein you eat to produce acid, but sugar!
You just cannot avoid sugar. Most of the food we eat contains "hidden" sugar. So, if you eat sweet food, the bacteria in your mouth metabolize the sugar and produce acid. The acid first attacks the dental enamel and then washes away the minerals inside your teeth. There should not be a problem having a sweet snack after a main meal, if you clean your teeth right afterwards. The worst thing you can do, however, is having frequent sweet snacks in between your meals. The dental enamel will then become permanently attacked. The acid contained in vinegar is harmful, too. That is why you should simply rinse your mouth with water after having any sweet or sour food containing sugar or vinegar, like sweets or salads for example, to neutralize the acid.

Tips for Healthy Teeth

- Clean your teeth after every meal. If this is not possible, rinse your mouth with water or use dental gum to neutralize the acid.
- Use sodium fluoride salt and mineral water. Sodium fluoride hardens the substance of the teeth. Green teas, legumes and fish contain fluoride, for example.
- Only use toothpaste containing fluoride.
- If you have eaten sour food like fruits or salads, rinse your mouth with water afterwards.
- Cheese is very good for the teeth. It replaces the minerals inside the dental enamel and neutralizes the acid in your mouth, too".

ACTIVITY 1

Find these words or expressions in the text and add them to your vocabulary list!

Ernährung	Flyer, ein Handzettel, ein Faltblatt	die Überschrift
sie soll lauten ...	ruinieren, kaputt machen	lediglich bestehend aus
Fett und Eiweiß	Kohlenhydrate	in Zucker verwandeln
Zucker meiden	der größte Teil der Nahrung	versteckter Zucker
den Zucker verstoffwechseln	zuerst, zunächst	Mineralien auswaschen
innerhalb der Zähne	Zwischenmahlzeit	Hauptmahlzeit
gleich nach, danach, hinterher	das Schlimmste	häufig
zwischen, dazwischen	ständig, fortlaufend, dauerhaft	Essig
ebenfalls schädlich	den Mund ausspülen	saure Nahrungsmittel
die Säure neutralisieren	Tipp	Natriumfluoridsalz
die (Zahn-) Substanz härten	Hülsenfrüchte, Gemüse	die Mineralien ersetzen

Healthy Food For Healthy Teeth!

UNIT 11

ACTIVITY 2

Read the text again and give an answer!

1. Fat and protein are turned into acid by the bacteria in your mouth. Is that true? Explain!
2. What happens with the sugar in your mouth?
3. What effect does the acid in your mouth have on your teeth?
4. Why is having a frequent sweet snack in between the meals very bad for your teeth?
5. What is a good thing to do after eating sweet or sour food?
6. Brushing your teeth after every meal very often is not possible. What could you do for your teeth even without having a toothbrush?
7. Why is it advisable to eat food and to use toothpaste containing sodium fluoride?
8. Why is eating cheese good for the teeth?

ACTIVITY 3

Say it in English!

1. Eiweiß und Fett verursachen Zahnbelag, aber greifen den Zahnschmelz nicht an.
2. Die Mundbakterien können aus Eiweiß und Fett keine Säure bilden.
3. Zucker aber ist der schlimmste Feind der Zähne!
4. Die Mundbakterien verstoffwechseln den Zucker und bilden (daraus) Säure.
5. Die Säure greift den Zahnschmelz an und wäscht die Mineralien in den Zähnen aus.
6. Das Schlimmste was man machen kann, ist häufig süße Sachen zu essen.
7. Der Zahnschmelz wird dann ständig von der Säure angegriffen.
8. Auch die im Essig enthaltene Säure ist schädlich und greift den Zahnschmelz an.
9. Sie können die Säure in Ihrem Mund neutralisieren. Spülen Sie Ihren Mund einfach mit Wasser aus.
10. Natriumfluorid ist sehr wichtig. Es härtet die Zahnsubstanz.
11. Deshalb sollten Sie nur Zahncreme verwenden, die Natriumfluorid enthält.
12. Käse ist gut für die Zähne.
13. Käse ersetzt die Mineralien, die sich im Zahnschmelz befinden, und neutralisiert auch die Säure im Mundraum.

False teeth

Man to woman: "Darling, your teeth remind me of the stars."
Woman: "Because they gleam and sparkle?"
Man: "No, because they come out at night!"

(false = falsch; remind = erinnern; stars = hier: Sterne; gleam = schimmern; sparkle = funkeln; to come out = herauskommen (Zähne), erscheinen (Sterne)

ACTIVITY 4

Work with a partner. Ask her or him questions!

1. Why is it important to use toothpaste containing sodium fluoride?
2. Why is sugar the worst enemy of healthy teeth?
3. What should you do after having a sweet snack?
4. Why is having a frequent sweet snack very bad for your teeth?
5. How can you neutralize the acid in your mouth after having eaten sweet or sour food?
6. Do you know foods containing fluoride?
7. Why is cheese good for the teeth?

ACTIVITY 5

Make a conversation with a partner!

Assistent	Patient
1. Dr. Zahn hat Sie gebeten, Ihre Patientin über den Zusammenhang von Ernährung und gesunden Zähnen aufzuklären. Sagen Sie Ihrer Patientin, dass Zucker der schlimmste Feind gesunder Zähne ist.	
	2. Sie möchten gerne wissen, wieso.
3. Sie erklären, warum.	
	4. Sie vermeiden zuviel Zucker ohnehin.
5. Sie erklären, warum das sehr schwer ist.	
	6. Statt süßer Sachen essen Sie lieber salzige Sachen und Salate.
7. Sie erklären die Wirkung von Essigsäure und wie man sie neutralisieren kann. Weisen Sie auf die Wichtigkeit von Fluorid hin.	
	8. Sie wissen das bereits.
9. Sie geben der Patientin den Tipp, Käse zu essen und erklären, warum.	
	10. Sie bedanken sich für die Ernährungstipps der Assistentin.
11. Sie erwidern den Dank.	

EXTENDED VOCABULARY LIST	
food	Essen, Nahrung, Lebensmittel, Nahrungsmittel
food poisoning	Lebensmittelvergiftung
dog food/cat food	Hundefutter/Katzenfutter
consist of	bestehen aus, sich zusammensetzen aus
fat	fett, dick, üppig, Fett
I got fat.	Ich bin dick geworden.
animal fat/vegetable fat	tierische(s) Fett(e)/Pflanzenfett(e)
carbohydrates, „carbs“	Kohlehydrate
metabolism	Stoffwechsel, Metabolismus
to metabolize	verstoffwechseln
wash away	auswaschen, ausspülen, wegspülen
meal	Mahl(zeit)/Essen, Schrotmehl
main meal	zu Hause: Hauptmahlzeit; Restaurant: main dish = Hauptgericht
right afterwards	gleich danach
Right over there.	Gleich da drüben.
Right opposite.	Gleich gegenüber.
Right now?	Sofort?, Gleich?
Right next to it.	Gleich daneben.
rinse	abspülen, ausspülen, tönen (Haar), die Spülung, Tönung
I'll rinse my hands.	Ich werde mir die Hände abspülen/waschen
frequent	häufig
frequency	Häufigkeit, Frequenz
vinegar	(Wein-) Essig
potassium	Kalium
sodium	Natrium

12 TOOTH DECORATION

Miss Sheila Kidman is an Irish girl of seventeen. She moved to Germany some time ago, because her father, Michael Kidman, started a job at an Irish bank branch here. Her German has improved a lot. However, if it comes to very special things, her German is not good enough yet. That is why she prefers to visit Dr. Zahn's dental surgery, because his entire staff speaks English. She booked a dental appointment as she would like to know something about tooth decoration.

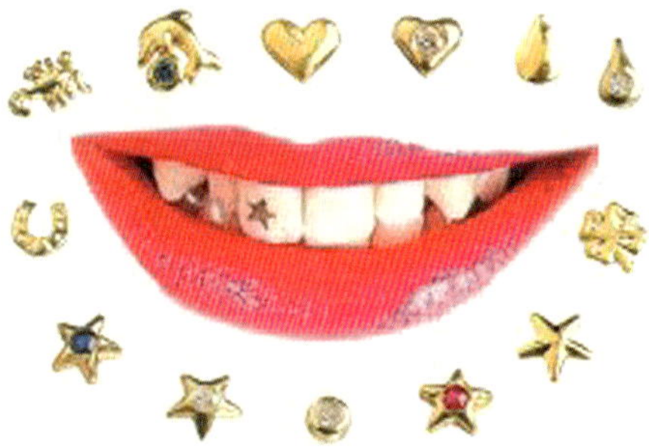

Sheila Hello.

Assistant Hello, Sheila. Could you please take a seat in our waiting room? Our hygienist is still busy. She will be with you in a few minutes. It won't take long.

8 minutes later

Assistant Sheila? Please go to treatment room 2. Our hygienist is waiting for you there.

In treatment room 2

Hygienist Hi, Sheila.

Sheila Hi.

Hygienist You came to get some information about tooth decoration?

Sheila Yes. I have been thinking about it for some time now and decided on having a tooth decoration on one of my front teeth.

Hygienist Well, tattoos, dazzlers and twinkles are real eye-catchers. No doubt about that. But you should know a few things before you decide having it done.

Sheila What's the difference between dazzlers and twinkles?

Hygienist Well, a dazzler is a tiny sheet of pure gold whereas a twinkle is a jewel. Those jewels can be diamonds, sapphires, rubies or gold. You can

	have real ones, if you can afford it, or just fashion jewellery. Some girls choose dazzlers. Dazzlers can hide an ugly spot on one of the teeth, too.
Sheila	I see. How do you fix them to the teeth?
Hygienist	Dazzlers and twinkles get bonded to the teeth. They are removable. That is, they can be taken off by the dentist any time.
Sheila	What about a tattoo? As far as I know, having something tattooed on the skin is very painful and if you want a tattoo removed, it's even more painful.
Hygienist	No, it isn't painful at all. Tattoos are mini-pictures. We bond them to the teeth as well. However, they may get off after some time.
Sheila	Well, I'd like to have a twinkle then.
Hygienist	OK. But, do not forget to be extra careful with your dental hygiene. Tooth decoration is an ideal hiding-place for bacteria.

ACTIVITY 1

Find these words or expressions in the text and add them to your vocabulary list!

ein 17-jähriges Mädchen	hier: umziehen, ziehen	vor einiger Zeit
eine Bankniederlassung	sich verbessern	noch nicht gut genug
etwas vorziehen, lieber tun	sein gesamtes Personal	Termin ausmachen (US)
Zahnverschönerung	noch beschäftigt sein	es wird nicht lange dauern
seit einiger Zeit	sich entscheiden	Zahnschmuck
eine Tätowierung, ein Tattoo	Goldblättchen/-folie	etwas, das „funkelt"
Blickfänger	zweifellos	der Unterschied zwischen
reines Gold	wohingegen	Schmuckstein, Juwel
Diamant	Saphir	Rubin
echte (z.B. Diamanten)	sich etwas leisten können	Modeschmuck
verbergen, verstecken	hässlich	Fleck, Punkt

Aha ...	etwas ankleben, verbinden mit	entfernbar
entfernen, abnehmen	jederzeit	soviel ich weiß
auf die Haut tätowieren	sehr schmerzhaft	Miniaturbilder
abgehen, abfallen	ich möchte gerne haben	dann
besonders sorgsam sein bei	Zahnhygiene, Zahnpflege	ideales Versteck

ACTIVITY 2

Read the text again and give an answer!

1. Sheila moved from Ireland to Germany. Why?
2. Why does she prefer visiting Dr. Zahn's dental surgery?
3. Why has she booked a dental appointment?
4. What kinds of tooth decoration do dentists offer?
5. Why do some female patients prefer having a dazzler?
6. How do dazzlers, twinkles and tattoos get fixed to the teeth?
7. Would it be a problem to take them off again?
8. Is getting a tattoo on the teeth painful?
9. Why should patients with tooth decoration be extra careful with their dental hygiene?

ACTIVITY 3

Say it in English!

1. Unser gesamtes Personal spricht Englisch.
2. Unsere Zahnhygienikerin hat noch etwas zu tun.
3. Sie wird in wenigen Minuten bei Ihnen sein.
4. Es wird nicht lange dauern.
5. Haben Sie sich jetzt für einen Twinkle entschieden?
6. Ein Twinkle ist ein echter Blickfänger.
7. Ein Twinkle ist ein kleiner Schmuckstein.
8. Oft ist es nur Modeschmuck.
9. Twinkles werden auf die Zähne geklebt.
10. Tattoos werden ebenfalls aufgeklebt.
11. Tattoos sind Minibilder.
12. Zahnschmuck ist ein ideales Versteck für Bakterien.
13. Deshalb sollte man besonders auf Zahnhygiene achten.

ACTIVITY 4

Work with a partner. Ask her or him questions!

1. What do you think about tooth decoration?
2. What kinds of tooth decoration are possible?
3. What is the difference between a dazzler and a twinkle?
4. Are twinkles always made of real jewels or gold?
5. How long will twinkles or dazzlers stick on the teeth?
6. Is getting a tattoo on the teeth painful?
7. Why should you be extra careful with your oral/dental hygiene, if you have a twinkle or a dazzler on one of your front teeth?

Tooth Tattoos

ACTIVITY 5

Make a conversation with a partner!

Assistant	Patient
1. Begrüßen Sie die Patientin.	
	2. Stellen Sie sich vor und sagen Sie, dass Sie um 9:30 Uhr einen Termin haben.
3. Bitten Sie die Patientin, im Wartezimmer Platz zu nehmen.	
	4. Fragen Sie, wo es sich befindet.
5. Erklären Sie, wo es ist.	

10 Minuten später

7. Rufen Sie die Patientin auf und bitten Sie sie, im Behandlungszimmer 1 Platz zu nehmen. Sagen Sie ihr, dass dort jemand auf sie wartet.	

Im Behandlungszimmer

8. Sie begrüßen die Patientin und fragen Sie sie, wie Sie ihr helfen können. Sagen Sie ihr, dass Dr. Zahn gleich kommen wird.

9. Sie hätten gerne einen Zahnschmuck. Fragen Sie, welche Möglichkeiten es gibt.

10. Zählen Sie auf, was es alles gibt.

11. Stellen Sie ein paar Fragen hierzu (z. B.: Wie wird Zahnschmuck befestigt, wie lange hält er? usw.)

12. Beantworten Sie die Fragen der Patientin.

13. Erwidern Sie etwas darauf.

14. Weisen Sie die Patientin auf die besondere Zahnhygiene bei Zahnschmuck hin.

15. Bedanken Sie sich für die Hinweise.

16. Erwidern Sie den Dank.

EXTENDED VOCABULARY LIST	
get bonded to	verbunden werden mit (z. B. Klebstoff)
branch	Ast, Zweig, Zweigstelle
dazzle	blenden, das Blenden, schimmern
decoration	Schmuck (zum Schmücken von irgendetwas), Dekoration
doubt	Zweifel
extra	zusätzlich, besonders, extra, Sonderleistung, Zuschlag
It is charged extra.	Es wird gesondert berechnet.
extras	Nebenausgaben, Nebeneinnahmen, Beilagen (Essen)
eye-catcher	Blickfang
fashion	Mode, Sitte, Brauch
in the usual fashion	wie üblich
it is in fashion	es ist modern, es ist in Mode
it is out of fashion	es ist unmodern, es ist aus der Mode
fashionable	schick

EXTENDED VOCABULARY LIST	
fix (to)	befestigen an, festmachen, aufhängen (Bild)
hiding-place	Versteck
jewel	Schmuckstein, Juwel
jewellery, auch: jewelry	Schmuck (Ringe, Kettchen, Perlen usw.)
not ... yet	noch ... nicht
prefer	bevorzugen, vorziehen, lieber haben als
preference	Vorzug, Vorliebe (for something)
sheet **sheet metal** **sheet of ice** **as white as sheet**	Blatt, Bogen, Betttuch, Scheibe Blech Eisfläche leichenblass
staff, personnel	die Angestellten, das Personal, die Belegschaft
stick **stick in** **stick on** **stick out**	kleben, steckenbleiben, haftenbleiben einkleben, einstecken haften-, kleben-, oben bleiben (Pferd) herausstehen (Nagel), abstehen (Ohren), auffallen
tattoo	Tätowierung, Tattoo, tätowieren, Militärmusikparade
tiny	winzig
twinkle	funkeln, glitzern, blitzen, das Funkeln
ugly	hässlich, grässlich, widerlich, abstoßend

13 TEETH GRINDING

Mrs. Nicole Benning, the American lecturer, has been working in Germany for three months now. Her new job is quite stressful and she has to work long hours. A week ago she visited a general practitioner because she suffers from painful jaws and headaches in the morning. The physician referred her to Dr. Zahn's dental surgery as he thought her pains might have something to do with her teeth ...

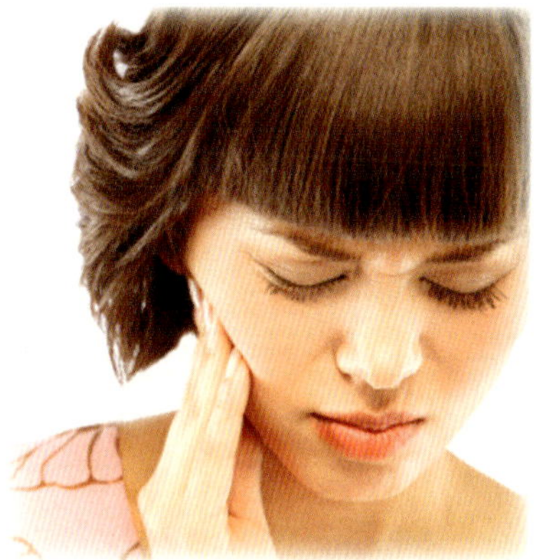

Mrs. Benning Good morning.

Assistant Good morning, Mrs. Benning.

Mrs. Benning I have an appointment for 9.30.

Assistant I know. Do you have a physician's letter with you?

Mrs. Benning Yes, I have. Here you are.

Assistant Great. Could you take a seat in our waiting room? I'll call you as soon as Dr. Zahn's ready for you ...

About 10 minutes later

Assistant Mrs. Benning? Could you please take a seat in room 1?

Mrs. Benning Excuse me, where is it?

Assistant Just down the hall. It's the second door on the left.

Mrs. Benning Thank you.

In room 1

Assistant Please, take a seat on the dental chair. Dr. Zahn will be with you in a minute.

Mrs. Benning Fine.

Assistant What seems to be the trouble, by the way?

Mrs. Benning Well, I suffer from severe neck pain, ear- and headache and my jaws feel tight. It is painful if I touch my cheeks. I went to see a physician. He said that grinding the teeth might be the cause of my troubles. A dentist may be able to help me, he said.

Assistant Did you know that some people grind their teeth with a biting force up to six times greater than normal?

Mrs. Benning Oh, really? I just can't believe it!

Assistant The dentists call this "Bruxism". Bruxism can cause the teeth to be ground down and become shortened. As the dental enamel is worn away by bruxism, the dentin layer gets exposed. The teeth become sensitive to cold, heat and pressure. The high pressure causes tooth pain and can even fracture teeth and crack fillings.

Mrs. Benning What can be done about it?

Assistant Well, Dr. Zahn can make a custom mouth guard. You wear it while you sleep. That's why we also call it "night guard". This can minimize the damage. However, it doesn't solve the problem. Ah, do you drink alcohol?

Mrs. Benning Not much. Why do you want to know?

Assistant Well, alcohol has been shown to worsen teeth grinding. It is also well known that mental problems, stress and anxiety play a major role in causing teeth grinding.

Mrs. Benning Really? I'm extremely stressed at the moment, you know.

Assistant You should do exercises to reduce muscle tension, relax and do meditation. If you successfully control your mental stress, you will dramatically reduce teeth grinding ...

A Mouth Guard

UNIT 13

ACTIVITY 1

Find these words or expressions in the text and add them to your vocabulary list!

Zähneknirschen	ziemlich stressig, anstrengend	lange Arbeitszeiten haben
aufsuchen	Facharzt für Allgemeinmedizin	schmerzende Kiefer (-gelenke)
Kopfschmerzen	Arzt	sie könnten vielleicht etwas damit zu tun haben mit ...

Arztbrief	Hier bitte!	Schön!, Toll!, Großartig!
den Gang entlang	Gut!	Welche Probleme haben Sie?
übrigens	leiden an	starke Nackenschmerzen
Ohrenschmerzen	Kopfschmerzen	hier: verspannt, hart, fest
berühren	die Wangen	knirschen, mahlen, zerreiben
es könnte sein	die Ursache	die Beißkraft
bis zu	sechs mal größer als	der Bruxismus (lat.), das Zähneknirschen
der Grund sein für etwas	abgeschliffen	verkürzt werden
hier: abnutzen, abtragen	die Dentinschicht	freigelegt, entblößt
Kälte, Hitze und Druck	brechen, zerbrechen	spalten, rissig werden lassen
etwas dagegen tun	angepasste/spezialangefertigte Knirscherschiene	tragen
während Sie schlafen	minimieren	den Schaden
er/sie/es hat sich erwiesen	das Zähneknirschen verschlimmern	es ist allseits bekannt
psychische Probleme	Angst, Ängste (psych.)	die Hauptrolle spielen bei
Übungen durchführen	vermindern	Muskelspannung
entspannen	etwas unter Kontrolle halten	psychischer/seelischer Stress, Anspannung

Patient: "Doctor, my teeth look so yellow. What do you recommend me to do?"

Dentist: "Wear a brown tie. That suits your yellow teeth best!"

(recommend = empfehlen; wear = tragen (Kleidung); tie = Krawatte; suit = passen zu)

ACTIVITY 2

Look at the text again and answer these questions!

1. Which symptoms has Mrs. Benning been suffering from?
2. Her physician referred her to Dr. Zahn's dental surgery. Why?
3. Her physician gave her a letter for Dr. Zahn to read. What kind of letter is it?
4. What are Mrs. Benning's troubles?
5. The biting force of the jaws is different with people suffering from bruxism. Explain!
6. What is "Bruxism"?
7. What can happen to your teeth if you suffer from it?
8. What can a dentist do about it?
9. What role can drinking alcohol play, if you suffer from bruxism?
10. What are the major reasons for teeth grinding?
11. What can a patient do about it?

ACTIVITY 3

Say it in English!

1. Meine neue Arbeit ist ziemlich stressig/anstrengend.
2. Meine Kiefergelenke schmerzen.
3. Ich denke, es hat etwas mit meinen Zähnen zu tun.
4. Haben Sie einen Arztbrief dabei?
5. Nehmen Sie bitte auf dem Untersuchungsstuhl Platz!
6. Entschuldigung, wo ist Zimmer 3?
7. Einfach den Gang entlang.
8. Zweite Tür links.
9. Dr. Zahn wird gleich zu Ihnen kommen.
10. Welche Probleme haben Sie?/Wie kann ich Ihnen helfen?
11. Frau Benning leidet an starken Nackenschmerzen, Ohren- und Kopfschmerzen.
12. Ihr Kiefergelenk fühlt sich verspannt an.
13. Tut es weh, wenn ich Ihre Wangen berühre?
14. Zähneknirschen könnte die Ursache Ihrer Probleme sein.

15. Zahnärzte bezeichnen Zähneknirschen als „Bruxismus“.
16. Zähneknirschen kann zur Folge haben, dass die Zähne abgeschliffen und kürzer werden.
17. Der Zahnschmelz wird abgetragen und die Dentinschicht freigelegt.
18. Die Zähne reagieren empfindlich auf Kaltes, Heißes und Druck.
19. Der hohe Druck kann sogar Füllungen zerbrechen.
20. Um den Schaden zu minimieren, können Zahnärzte individuelle Knirscherschienen anfertigen.
21. Alkohol kann das Problem verschlimmern.
22. Die Hauptursachen sind seelische Probleme, Stress und Ängste.
23. Wenn es einem erfolgreich gelingt, seelische Probleme in den Griff zu bekommen, kann man Zähneknirschen erheblich reduzieren.

>>> ACTIVITY 4 >>>

Work with a partner. Ask her or him questions!

1. Which are the major causes for bruxism?
2. What can worsen the problem?
3. What would you advise a patient to do about it?
4. What can a dentist do about it?
5. Which are the typical symptoms that people show who suffer from teeth grinding?
6. What do you know about the biting force of jaws?

>>> ACTIVITY 5 >>>

All these statements are false. Look at the text again and correct them!

1. Bruxism causes the teeth to become yellow and to build up tartar.
2. The major cause of teeth grinding is drinking alcohol.
3. People who grind their teeth during sleep feel dizzy and sick.
4. By making custom mouth guards, dentists are able to solve all the problems which cause teeth grinding.
5. The biting force of the jaws with people suffering from bruxism is ten times greater than with other people.
6. Reducing mental stress has no effect on teeth grinding.
7. The biting force is too weak to crack fillings or fracture teeth.

ACTIVITY 6

Make a conversation with a partner!

Assistant	Patient
	1. Sie betreten die Praxis, grüßen, stellen sich vor und sagen, dass Sie Ihr Hausarzt überwiesen hat. Ihr Termin ist um ... Uhr. Sie haben einen Arztbrief dabei.
2. Sie bestätigen den Termin, lassen sich den Arztbrief geben und bitten die Patientin, kurz im Wartezimmer Platz zu nehmen. Sie wird in etwa 5 Minuten aufgerufen werden.	
	3. Sie erwidern etwas darauf.
4. Sie rufen die Patientin auf und bitten sie, im Behandlungszimmer Platz zu nehmen.	
	5. Sie fragen, wo es sich befindet.
6. Sie erklären kurz den Weg und sagen, dass sie dort von einer Kollegin erwartet wird.	

Im Behandlungsraum

Assistant	Patient
7. Sie sind jetzt die Kollegin. Sie bitten die Patientin, auf dem Untersuchungsstuhl Platz zu nehmen. Fragen Sie nach ihrem Problem.	
	8. Beschreiben Sie es: Nacken-, Ohren- und Kopfschmerzen, verspanntes und schmerzendes Kiefergelenk, Berührungsschmerz an den Wangen. Sagen Sie, was Ihr Hausarzt als Ursache vermutet.
9. Bestätigen Sie die Vermutung des Hausarztes der Patientin. Erzählen Sie der Patientin, was Sie über Bruxismus wissen und wie sich dies letztlich auf die Zähne auswirken kann.	
	10. Fragen Sie, was ein Zahnarzt in so einem Falle tun kann.
11. Erwähnen Sie die Möglichkeit der Anfertigung einer individuell angefertigten Knirscherschiene. Vergessen Sie nicht darauf hinzuweisen, dass man damit zwar die Schäden für die Zähne begrenzen, aber nicht die Ursache beheben kann.	

12. Fragen Sie nach den Ursachen für Zähneknirschen.

13. Zählen Sie mögliche Ursachen auf.

14. Erwähnen Sie, dass Sie zurzeit einem enormen Stress ausgesetzt sind.

15. Machen Sie Vorschläge, wie die Patientin von sich aus ihr Problem in den Griff bekommen könnte.

EXTENDED VOCABULARY LIST	
anxiety	Angstgefühl, Besorgnis
by the way	übrigens
Sit by me.	Setz dich zu mir. Setzen Sie sich zu mir.
I walked by the house.	Ich bin am Haus vorbeigegangen.
by day	am Tag, tagsüber
by tomorrow	bis morgen
by now	bis jetzt, bisher
By then!	Bis dann!
by myself	ganz allein, ohne Hilfe
little by little	allmählich, nach und nach, langsam, schrittweise
one by one	einer nach dem anderen
by chance	zufällig
by the dozen	dutzendweise
by no means	auf keinen Fall
by nature	von Natur aus
I can't get by.	Ich komme nicht vorbei.
stand by	bereit sein, in der Nähe sein
by and by	nach und nach
by and large	im Großen und Ganzen
corridor (UK), hall (US)	Korridor, Hausflur
crack	springen, bersten, brechen, Sprung, Riss, Spalt
custom	Sitte, Brauch, speziell, Maß-
custom-made	maßangefertigt, kundengerecht, individuell angefertigt
custom-built	spezialangefertigt
custom clothes	Maßkleidung
customer	Kunde

EXTENDED VOCABULARY LIST	
customs	Zoll
dentin	Zahnbein
dizzy	schwindelig, benommen
do about	etwas tun gegen ...
expose	freilegen, entblößen, aufdecken, enthüllen
force	Druck, Kraft, Stärke, Zwang, Gewalt, Macht
by force	mit Gewalt
in full force	in voller Stärke, vollzählig
force of character	Charakterstärke
the force of habit	die Macht der Gewohnheit
fracture	Fraktur, Bruch, brechen
general practitioner	Allgemeinarzt (oft auch family doctor = Hausarzt)
grind/ground/ground	knirschen (Zähne), (zer)mahlen, schleifen, wetzen
grind down	abschleifen, abwetzen
guard	Schutz, Wache, Wärter, bewachen, schützen
jaw	Kiefer, Kiefergelenk, Maul (Haifisch)
layer	Schicht
major role	Hauptrolle
major operation	eine größere Operation
mental	geistig, seelisch, mental
mental blackout	Bewusstseinsstörung
mental breakdown	Nervenzusammenbruch
mental health	seelischer Zustand
mental hospital	Nervenklinik
mental home	Nervenheilanstalt
mental illness	Geisteskrankheit
minimize	minimieren, auf ein Minimum reduzieren
mouth guard, also: night guard	hier: Knirscherschiene als Zahnschutz
painful	schmerzhaft
physician	Arzt

EXTENDED VOCABULARY LIST	
quiet	ruhig, still
quite	ziemlich, wirklich, völlig
refer to, refer/referred/referred	überweisen an (Patient), verweisen, anweisen, beziehen
relax	lockern, entspannen, nachlassen, sich erholen
sensible	vernünftig
sensitive (to)	sensibel, empfindlich (auf)
severe **severe test** **severe storm** **severe weather**	schlimm, schwer (Krankheit) schwere Prüfung heftiger Sturm raues Wetter
sick	übel, krank
stressful	stressig, anstrengend
tight	fest, verspannt, enganliegend, knapp (Geld), geizig, besoffen
wake (wake/woke/woken)	erwachen, munter werden, wach, wecken
wear away	auswaschen, abtragen (Kleidung, Schuhe), aushöhlen
worse (bad/worse/worst)	schlimmer, schlechter
worsen	verschlimmern, verschlechtern

14 THE IMPORTANCE OF FLOSSING

Mrs. Savannah Williams, another English speaking patient of Dr. Zahn's dental surgery, has had professional dental cleaning. After the treatment, Barbara, Dr. Zahn's dental hygienist, gives Mrs. Williams some advice on how to improve her dental care. She recommends Mrs. Williams to regularly floss her teeth. Mrs. Williams already brushes her teeth twice a day. Sometimes even after meals. So, she does not think it is necessary to floss. However, the dental hygienist tries to convince her to do it all the same ...

Barbara Next to tooth brushing, flossing is the most important thing to get healthy teeth.

Mrs. Williams I already brush my teeth several times a day. So, why should I floss them?

Barbara Flossing removes the bacteria which hide in the tiny spaces in between your teeth. Brushing without flossing is like washing only 65 per cent of your body! I already told you that the bacteria feed on food particles left in your mouth. And it is the acid they produce which eats into the dental enamel, washing away the calcium and creating cavities.

Dental Floss

Mrs. Williams I use a mouthwash. I thought it would kill all the bacteria in my mouth.

Barbara Well, this effect does not last. Twenty minutes later, most of the bacteria are back again. However, the bacteria won't have a chance to produce acid, if you brush your teeth **and** remove all the food particles and the plaque in between your teeth. If you never floss, the plaque hardens into calculus. Unlike plaque, which can be easily removed by brushing, calculus, that is tartar, can only be removed by a dentist.

Mrs. Williams I see ...

Barbara You know, if the plaque has hardened into tartar, even more dangerous bacteria build up within the tartar. The bacteria not only produce acid which attacks the enamel but also toxins which inflame the gums.

This is known as gingivitis. If the gingivitis is left untreated, you may get periodontal disease. This disease affects the structures supporting your teeth. Your teeth become loose and fall out. The bacteria also pour out sulphur compounds creating embarrassing bad breath.

Mrs. Williams I have heard that they sell different kinds of dental floss. Which one should I use then?

Barbara I'd recommend you to use waxed dental floss because it slides more easily between your teeth than normal one. However, you can use any other floss as well, of course.

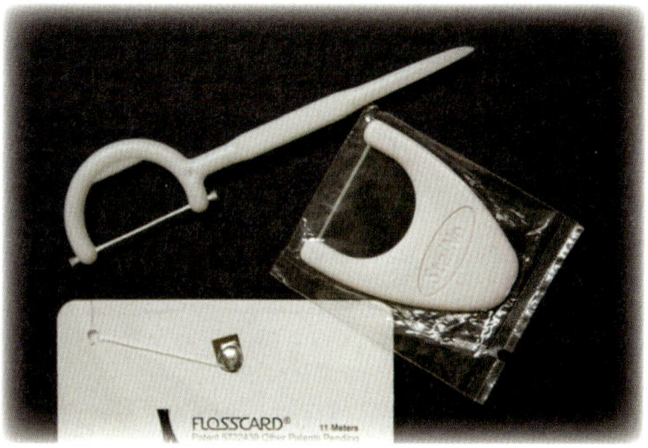

ACTIVITY 1

Find the words or expressions in the text and add them to your vocabulary list!

die Wichtigkeit	das Benutzen von Zahnseide	Rat geben hinsichtlich
die Zahnpflege verbessern	sie versucht	zu überzeugen
trotzdem	gleich nach dem Zähneputzen	die winzigen Zahnzwischenräume
Prozent	sich ernähren von	Essensreste
übrig, übriggeblieben	sich hineinfressen	Zahnschmelz
Löcher bilden/schaffen	es dauert nicht an	es verhärtet sich zu Zahnstein
im Gegensatz zu Zahnbelag	leicht entfernen	Bakterien entstehen
innerhalb des Zahnsteins	Toxine, Giftstoffe	Zahnfleischentzündung
sie bleibt unbehandelt	Erkrankung des Periodonts	befallen, angreifen, in Mitleidenschaft ziehen
der Zahnhalteapparat	locker werden	ausfallen
ausströmen, ausschütten	Schwefelverbindungen	schaffen, entstehen lassen
peinlich, unangenehm	schlechter Atem	gewachste Zahnseide
leichter	jede andere Zahnseide	ebenso

ACTIVITY 2

Read the text again and give an answer!

1. Which effect has flossing got?
2. What do the bacteria in your mouth feed on?
3. The bacteria produce acid. What does the acid do to your teeth?
4. What effect does a mouthwash have?
5. You should not give the bacteria a chance to produce acid. What can you do to avoid the production of acid in your mouth?
6. Can you remove calculus/tartar by brushing or flossing?
7. What will happen, if you do not remove plaque?
8. The bacteria in your mouth not only produce acid. What else do they produce?
9. What can happen, if gingivitis is left untreated?
10. There are some people who suffer from embarrassing bad breath. What might be one reason for having bad breath?
11. Why is it better to use waxed dental floss?

What can you see in the picture?

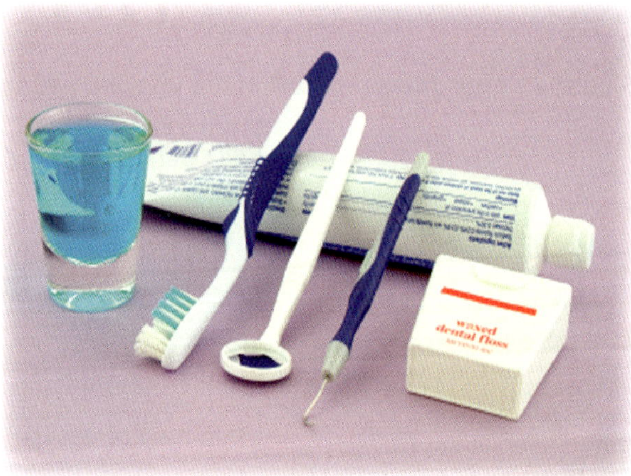

1. _ O _ _ _ _ _ _ S H
2. T _ _ _ H _ _ _ _ _ _
3. D _ _ _ _ _ _ _ I _ K
4. _ O _ _ _ _ _ _ _ _ E
5. _ _ _ _ _ _ L _ _ R _ _ _
6. _ A _ _ _ D _ _ _ _ _ _ _ _ _ _ S

ACTIVITY 3

Say it in English!

1. Durch die Verwendung von Zahnseide können Sie die Bakterien entfernen, die sich in den winzigen Zahnzwischenräumen verbergen.
2. Die Bakterien ernähren sich von Nahrungsresten, die im Mund verbleiben.
3. Die Säure frisst sich in den Zahnschmelz, wäscht das Kalzium aus und bildet Löcher.

4. Eine Mundspülung kann die Mundbakterien abtöten.
5. Die Wirkung hält jedoch nicht an.
6. Wenn Sie nicht nur Ihre Zähne reinigen, sondern auch alle Nahrungsreste **und** den Zahnbelag zwischen Ihren Zähnen entfernen, haben die Bakterien keine Möglichkeit, Säure zu produzieren.
7. Wenn Sie nie Zahnseide verwenden, verhärtet sich der Zahnbelag zwischen den Zähnen zu Zahnstein.
8. Zahnbelag kann leicht durch Bürsten oder Zahnseide entfernt werden.
9. Zahnstein kann nur von einem Zahnarzt oder einem/einer Dental-Hygieniker/in entfernt werden.
10. Im Zahnstein entstehen sogar noch gefährlichere Bakterien.
11. Die Bakterien produzieren Säure und Toxine, welche das Zahnfleisch entzünden.
12. Bleibt eine Zahnfleischentzündung unbehandelt, kann man eine Parodontitis bekommen.
13. Dies ist eine Krankheit, welche den Zahnhalteapparat in Mitleidenschaft zieht/zerstört.
14. Ihre Zähne werden locker und fallen aus.
15. Die Bakterien produzieren auch Schwefelverbindungen, die peinlich schlechten Atem verursachen.
16. Ich würde Ihnen empfehlen, gewachste Zahnseide zu verwenden.
17. Gewachste Zahnseide gleitet besser zwischen Ihren Zähnen.

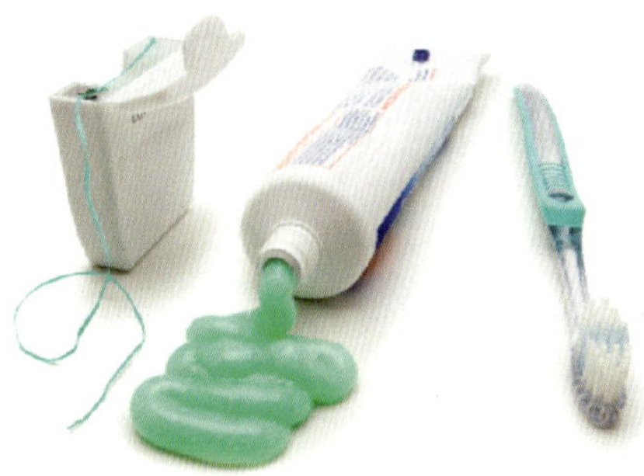

Patient: "Every time I eat bananas, I get this strange urge to give people all my money".
Dentist: "Do you want a banana?"

(a strange urge = ein merkwürdiger Drang)

UNIT 14

ACTIVITY 4

Ask your partner a few questions!

1. What advice would you give a patient how to improve her or his dental care?
2. Why should you recommend a patient to regularly floss her or his teeth?
3. How can a patient keep the bacteria in the mouth from producing acid?
4. What happens, if the plaque is not regularly removed by brushing and flossing?
5. What is "periodontal disease"?
6. What might be one reason for having bad breath?

ACTIVITY 5

Make a conversation with a partner!

Assistant	Patient
1. Betonen Sie gegenüber der Patientin, wie wichtig es ist, auch Zahnseide zu verwenden.	
	2. Sie sehen den Grund nicht ein, denn Sie reinigen Ihre Zähne ohnehin bereits mehrmals täglich. Manchmal nach jeder Mahlzeit.
3. Erklären Sie, warum Zähneputzen nicht ausreicht.	
	4. Sie verwenden auch eine Mundspülung. Für eine gute Zahnpflege sollte das alles eigentlich ausreichen, meinen Sie.
5. Eine Mundspülung hat keinen nachhaltigen Effekt. Erklären Sie dies der Patientin. Erklären Sie auch, warum der Zahnbelag in den Zwischenräumen unbedingt regelmäßig entfernt werden muss.	
	6. Sie haben eine leichte (=mild) Zahnfleischentzündung. Sie fragen, ob diese durch die Bakterien im Zahnstein verursacht sein könnte.
7. Sie bejahen dies und erklären den Zusammenhang zwischen Zahnstein, Bakterien, Toxinen und einer Zahnfleischentzündung. Weisen Sie auf die Folgen einer unbehandelten Zahnfleischentzündung hin.	
	8. Sie haben manchmal schlechten Atem. Fragen Sie nach den Gründen.
9. Erklären Sie der Patientin die möglichen Gründe.	
	10. Sie sehen jetzt ein, dass die Verwendung von Zahnseide durchaus sinnvoll ist. Fragen Sie die Helferin, welche Zahnseide sie empfehlen würde.
11. Beraten Sie die Patientin.	

UNIT 14

EXTENDED VOCABULARY LIST

affect (verb)	beeinträchtigen, in Mitleidenschaft ziehen, befallen
all the same	dennoch, trotzdem
another	eine/ein weitere/r
convince	überzeugen

EXTENDED VOCABULARY LIST	
create	schaffen, hervorbringen, verursachen, bewirken, gründen
eat into	hineinfressen
embarrassing **to embarrass someone**	peinlich, unangenehm jemanden in Verlegenheit bringen
feed on something	sich von etwas ernähren
gingiva	Zahnfleisch (lat.)
gum/gums **bleeding of the gums**	Zahnfleisch Zahnfleischbluten
harden **become hardened (to)**	härten, abbinden, stählen, stärken, festigen abgehärtet werden (gegen)
harden into something	sich in etwas verhärten
to keep from producing something	davon abhalten, etwas zu produzieren, hier: Säure
last (adjective) **for he last time** **last night** **I'm the last but one.** **the week before last** **That's the last thing I would do.** **the last word in fashion**	letzte/r/s, neueste/r/s, vorig/vergangen zum letzten Mal gestern Abend, vergangene Nacht Ich bin der Vorletzte vorletzte Woche Das wäre das Letzte was ich täte. das Neueste in der Mode, der „letzte Schrei“
to last (verb)	andauern, anhalten, aushalten, ausreichen
left	übrig, links
loose (adjective)	locker, lose, frei, weit (Kleidung)
lose/lost/lost (verb)	verlieren/verlor/verloren
part	Teil, Stück eines Ganzen, Anteil, Abschnitt, Theaterrolle
particle **not a particle of sense**	Teilchen, Partikel kein Fünkchen Verstand
per cent	Prozent
periodontal disease	Parodontitis, Erkrankung des Zahnhalteapparates, früher auch als Parodontose bezeichnet

EXTENDED VOCABULARY LIST	
pour	gießen, schütten
pour out	ausströmen, ausschütten, ausgießen
pouring rain	strömender Regen
several	einige, mehrere, verschiedene
several of us	einige von uns
We went our several ways.	Wir gingen getrennte Wege
several times	mehrere Male
structure	Struktur, Aufbau, Gefüge, Konstruktion
sulphur	Schwefel
sulphur dioxide	Schwefeldioxid
sulphurous	schwefelhaltig
sulphur compounds	Schwefelverbindungen, Schwefelteilchen
support	stützen, tragen, unterstützen, fördern, Unterstützung
surgery	die Chirurgie, das Sprechzimmer, die Praxis
I'll have surgery.	Ich werde operiert.
surgery hours	Sprechstunden (Zeiten)
surgical	chirgurgisch
surgical ward	die chirurgische Abteilung (einer Klinik)
toxic	giftig
toxic agent	Giftstoff
toxic waste	Giftmüll
toxin	Gift, Toxin
treat	behandeln
untreated	unbehandelt
It is left untreated.	Es bleibt unbehandelt.

15 PERIODONTAL DISEASE

Mr. Vito Corleone speaks both Italian and English but very little German. In Malta, he worked as an engineer for a German producer of children's car seats. As the German company shifted its production back to Germany, Mr. Corleone moved to Germany, too. His teeth are in a very bad condition. He is 48 years old. He has lost a few teeth already and some of his front teeth are loose. He visits Dr. Zahn's dental surgery to get help ...

Mr. Corleone Guten Morgen.

Assistant Guten Morgen. Was kann ich für Sie tun?

Mr. Corleone Excuse me, do you speak Italian?

Assistant No, sir.

Mr. Corleone Do you speak English?

Assistant Yes, a little. What can I do for you?

Mr. Corleone My name is Corleone. Vito Corleone. I've booked an appointment for 11.45.

Assistant Corleone? Ah, yes. Could you please put your personal details on this form here and take a seat in the waiting room? We are a bit busy because we had two emergencies this morning, you know. So, it will take some time until Dr. Zahn can see you.

30 minutes later

Assistant Mr. Corleone? Could you please come along to the examination room? Our dental hygienist will take care of you until Dr. Zahn is ready for you.

In the dental room

Hygienist Hello, I'm Barbara. I'm the dental hygienist. Please take a seat on the dental chair here. I'd like to have a look at your teeth. Please, open up nice and wide and turn your head towards me ... Yes, that's great.
Oh, that doesn't look good at all.

Mr. Corleone Yes, I know. That's why I came here.

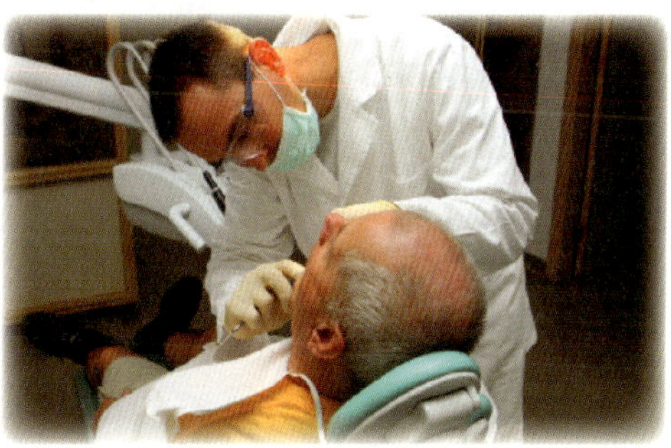

Hygienist Well, your gums are swollen and red, these two front teeth are loose and there's pus coming from around this molar here. Do you suffer from pain on chewing?

Mr. Corleone Oh yes, I do. My gums bleed when I brush my teeth. My gums are extremely tender and my teeth are very sensitive to cold, hot, salty and sweet things, too.

Hygienist The necks of your teeth are exposed. That makes the cervical nerve-endings vulnerable. It looks like you are suffering from a severe periodontal disease.

Mr. Corleone What is that?

Hygienist Well, it all starts with plaque. If not removed within 48 hours, plaque hardens into tartar. Once tartar develops, the only way to remove it is by having the teeth professionally cleaned. Tartar, however, is the home for millions of very harmful bacteria. Their toxins cause gum infection. This activates the immune system and results in an inflammation of the gums. If only the gums are affected, the disease is called *gingivitis*. If the bones and ligaments which hold the teeth are affected, it is called *periodontitis*. This disease is caused by a bacterial infection of the structures which hold the teeth.

Mr. Corleone That sounds complicated ...

Assistant The periodontal pockets, that are the spaces between your gums and your teeth, are much deeper than 3 millimetres. I'm afraid, your periodontal disease will require some dental surgery. Well, first we will have to make an X-ray of your teeth ...

Mr. Corleone Is it that bad?

Assistant I'm afraid, yes. The best advice I can give you for the future is brushing and flossing to remove plaque and seeing your dentist every six month to remove tartar that might build up. It is advisable to use special toothpastes for gum diseases, like bleeding gums and periodontal disease. There are several products on the market.

Mr. Corleone I'm afraid, you are right there ...

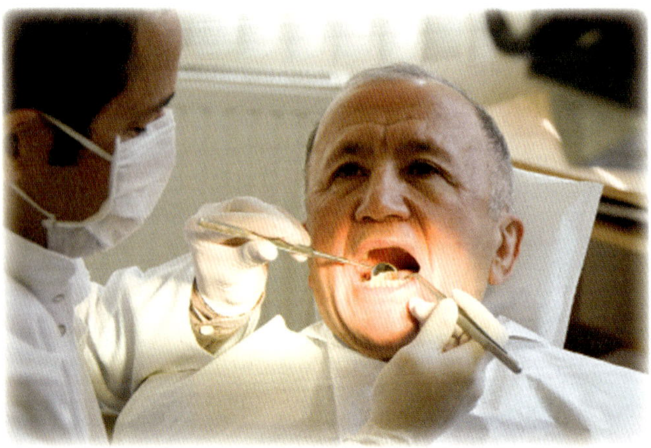

ACTIVITY 1

Find these words or expressions in the text and add them to your vocabulary list!

Erkrankung des Periodonts hier: Parodontitis	sowohl Deutsch als auch Englisch	wenig Deutsch sprechen
ein Ingenieur	ein Hersteller	Autokindersitze
die Produktion verlagern	nach Deutschland ziehen	in einem schlimmen Zustand
locker	persönliche Daten/Angaben	Notfälle
mitkommen	sich kümmern um	einen Blick werfen auf
schön weit aufmachen	geschwollen	es kommt Eiter heraus
Backenzahn, Molar	Schmerzen beim Kauen	extrem (schmerz-)empfindlich
kälte- u. hitzeempfindlich	Zahnhälse	freiliegend
ungeschützt	eine schlimme Krankheit	entfernt
innerhalb	zu Zahnstein verhärten	schädliche Bakterien
Toxine, Giftstoffe	Infektion des Zahnfleisches	aktivieren, in Gang setzen
das Immunsystem	es resultiert in	Zahnfleischentzündung
betroffen, erkrankt, angegriffen	Gingivitis	Knochen und Bänder
Parodontitis	Gewebe	es klingt kompliziert
die Zahntaschen	die Räume zwischen	einen zahnchirurgischen Eingriff erfordern
Ist es so schlimm?	Ich fürchte, ja.	für die Zukunft
den Zahnarzt aufsuchen	es könnte entstehen	Da haben Sie recht!

ACTIVITY 2

Look at the text again and give an answer!

1. What is wrong with Mr. Corleone's teeth?
2. What kind of dental disease is he suffering from?
3. Which languages does he speak?
4. He is Italian and worked in Malta. Why has he moved to Germany?
5. Why will he have to wait for some time before Dr. Zahn can see him?
6. Before Dr. Zahn comes to see him, the dental assistant has a look at Mr. Corleone's teeth. What does she see?

7. Mr. Corleone tells the hygienist what trouble he has got with his teeth. Describe his dental problems!
8. Describe how negligent dental hygiene may cause periodontitis!
9. What is the cause of a periodontal disease?
10. What are "periodontal pockets"?
11. What advice does the hygienist give the patient for the future?

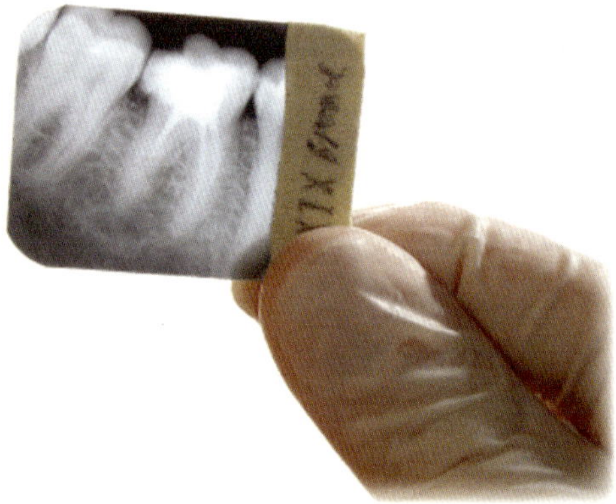

Dental X-ray

ACTIVITY 3

Say it in English!

1. Herr Corleone spricht sowohl Italienisch als auch Englisch.
2. Er hat für 10:30 Uhr einen Zahnarzttermin ausgemacht.
3. Könnten Sie auf diesem Formular Ihre persönlichen Daten angeben?
4. Unsere Praxis ist heute ziemlich voll.
5. Wir hatten heute Morgen drei Notfälle.
6. Es wird etwas dauern.
7. Könnten Sie bitte mitkommen?
8. Eine Helferin wird sich um Sie kümmern.
9. Hallo, ich bin Barbara. Ich bin die Dental-Hygienikerin.
10. Nehmen Sie bitte auf dem Untersuchungs/Behandlungsstuhl Platz!
11. Ich möchte mir gerne Ihre Zähne ansehen.
12. Machen Sie Ihren Mund schön weit auf!
13. Wenden Sie bitte Ihren Kopf in meine Richtung!
14. Oh, das sieht (aber) gar nicht gut aus.
15. Dieser Schneidezahn ist locker.
16. Ihr Zahnfleisch ist geschwollen und rot.
17. Haben Sie Schmerzen beim Kauen?
18. An diesem Backenzahn kommt Eiter heraus.
19. Mein Zahnfleisch ist extrem (schmerz-) empfindlich.
20. Meine Zähne sind sehr kälte- und wärmeempfindlich.
21. Sie haben eine starke/schlimme Parodontitis.
22. Aus Zahnbelag wird Zahnstein, wenn er nie entfernt wird.
23. Im Zahnstein leben Millionen schädlicher Bakterien.
24. Ihre Gifte verursachen eine Infektion des Zahnfleisches.
25. Die Infektion aktiviert das Immunsystem.
26. Das Ergebnis ist eine Zahnfleischentzündung, auch Gingivitis genannt.
27. Wenn man nichts dagegen unternimmt, schreitet die Gingivitis fort und es kann eine Parodontitis daraus entstehen.

28. Parodontitis wird durch eine bakterielle Infektion des Knochens und der Bänder verursacht, welche die Zähne halten.
29. Die Zahnfleischtasche ist der Raum zwischen dem Zahnfleisch und dem Zahn.
30. Bei einer Parodontitis ist häufig/oft ein chirurgischer Eingriff notwendig.
31. Ihre Zahnhälse liegen frei.
32. Deshalb sind Ihre Zahnnerven ungeschützt.

>>> ACTIVITY 4 >>>

Ask your partner a few questions!

1. Do you know what a *periodontal disease* is?
2. What may happen, if you do not remove plaque within 48 hours?
3. Have you ever suffered from pain on chewing?
4. Do your gums bleed sometimes?
5. Are your teeth sensitive to hot and cold?
6. Do you have many patients suffering from periodontal disease?
7. How can you get a gum infection?
8. An infection activates the immune system. How does the immune system react to an infection?
9. What is a *gingivitis*?
10. How do you call the disease infecting the bone and ligaments which hold the teeth in English and German?

>>> ACTIVITY 5 >>>

Make a conversation with a partner!

Patient	Assistant
1. Sie betreten die Praxis, grüßen und stellen sich vor. Sie haben um 9:45 Uhr einen Termin.	
	2. Sie begrüßen die Patientin. Sie bestätigen den Termin, weisen aber auf eine Verzögerung wegen eines Notfalles hin.
3. Sie fragen, wie lange Sie voraussichtlich warten müssen.	
	4. Sie nennen eine Zeit. Bitten Sie die Patientin, ein Formular mit persönlichen Angaben auszufüllen und danach im Wartezimmer Platz zu nehmen.

Einige Zeit später

5. Rufen Sie die Patientin auf und bitten Sie sie, auf dem Behandlungsstuhl Platz zu nehmen. Sie würden sich gerne die Zähne der Patientin ansehen und bitten sie daher, ihren Mund möglichst weit zu öffnen und ihren Kopf in Ihre Richtung zu drehen ...
Die Zähne der Patientin sehen nicht gut aus. Beschreiben Sie, was Sie sehen: lockere Zähne, Eiter, geschwollenes und entzündetes Zahnfleisch, Zahnstein, freiliegende Zahnhälse. Sie stellen fest, die Patientin hat Parodontitis.

6. Sie möchten wissen, was eine *Parodontitis* genau ist.

7. Erklären Sie es.

8. Beschreiben Sie, welche Probleme Sie haben: das Kauen schmerzt, die Zähne sind hitze- und kälteempfindlich, und das Zahnfleisch blutet beim Putzen.

9. Beschreiben Sie, wie sich aus nicht entferntem Zahnbelag zuerst Zahnstein, dann eine Zahnfleischentzündung und schließlich eine Parodontitis entwickeln kann.

10. Sie fragen, was jetzt noch getan werden kann.

11. Sie weisen darauf hin, dass vermutlich zahnchirurgische Eingriffe nötig sein werden.

12. Sie möchten wissen, was Sie tun können, damit es künftig nicht mehr so weit kommen wird.

13. Geben Sie Ratschläge zur künftigen Zahnhygiene der Patientin.

EXTENDED VOCABULARY LIST

activate	aktivieren
advisable	ratsam
dentin(e)	Zahnbein, Dentin
cervical	am Zahnhals
both ... and	sowohl ... als auch
children's car seat	Kindersitz
dental (treatment) chair	(Zahn-) Behandlungsstuhl
dental surgery	Zahnchirurgie, Zahnarztpraxis
emergency	Notfall
emergency room	Notaufnahme eines Krankenhauses
engineer	Ingenieur, Ingenieurin, Techniker

EXTENDED VOCABULARY LIST	
exposed	freiliegend
inflamed	entzündet
inflammation	Entzündung
image	Bild, Abbildung
ligament	Band, Ligament
loose	locker
lose/lost/lost	verlieren/verlor/verloren
pain on chewing	Schmerz(en) beim Kauen
periodontal disease, periodontitis	Erkrankung des Periodonts, Parodontitis, bakterielle Infektion des Zahnhalteapparates, veralteter Begriff: Parodontose
periodontal pockets	Zahntaschen
pus	Eiter
require	erfordern
result in	als Ergebnis haben, resultieren in
severe disease	schlimme Krankheit
shift **night shift**	verlegen, Schicht Nachtschicht
state	der Zustand, der Staat
tender	schmerzempfindlich
that's why ...	deswegen, deshalb, daher
tooth neck	Zahnhals
tooth nerve	Zahnnerv
turn into	sich verwandeln in
unprotected	ungeschützt

A "Dental" Poem

My dentist plans new teeth for me,
He'll fix them on the spot,
The only problem is, they'll cost,
More money than I've got.

(poem = Gedicht; fix = befestigen, anbringen, einsetzen; on the spot = auf der Stelle, sofort)

16 SENSITIVE TEETH

Once in a while, Dr. Zahn talks with his assistants about dental problems. Sometimes, he does that in English, just to practice the most important foreign language. Today he is talking about "What to do about sensitive teeth?" As everyone knows, sensitive teeth can make life miserable. They cause pain when eating cold or hot food, having cold or hot drinks or breathing cold air. He tells his assistants, that all these symptoms are signs of unprotected dental nerves.

Dr. Zahn Let's talk about sensitive teeth today. Well, it's important to tell our patients that sensitive teeth can be successfully treated. By the way, what's the Latin term for "tooth sensitivity"?

Assistant A In Latin, it's called "dentin hypersensitivity".

Dr. Zahn Right. How does the 'problem start?

Assistant B Well, it all starts with gingival recession.

Dr. Zahn Right. Normally, the gums cover the necks of the teeth to prevent the dental nerves from causing pain. For which reasons do gums recede?

Assistant B Many people brush their teeth with too much pressure. Using an electric toothbrush is better. People use less pressure when using electric toothbrushes than when using their hands. Also, many people use highly abrasive toothpastes which destroy the dental enamel, making teeth sensitive to heat, cold and sweet things. Dental enamel is the hardest substance in the human body. The white enamel covers the yellowish dentin and protects the much softer dentine from decaying quickly. However, over weeks, month and years of brushing and using abrasive toothpastes, people not only remove the gum tissue and expose the dental nerves, they also "brush off" the enamel. Teeth without enamel look yellow. Even bleaching cannot make them white again. You can only bleach the enamel but not the dentin.

Dr. Zahn Very good. Too much pressure when brushing the teeth is no good as well as using highly abrasive toothpastes. Are there any more reasons?

Assistant A Many people do not brush their teeth properly. So, plaque builds up. Over time, the plaque hardens into calculus. The bacteria in the calculus are responsible for the gingivitis and periodontitis that cause gum recession.

Dr. Zahn What would you advise the patients to do to prevent their gums from receding?

Assistant A Proper oral hygiene and using mildly abrasive or even better, non-abrasive toothpastes would help a lot.

Dr. Zahn Which dental hygiene tips would you give the patients?

Assistant B They should brush their teeth at least for 2 minutes and use floss to reach the tooth surfaces brushing cannot reach.

Dr. Zahn Good. You know, many people spend less than 45 seconds brushing their teeth. OK. What can patients do, if they already have sensitive teeth?

Assistant B I would advise the patients to use non-abrasive or mildly abrasive sensitivity toothpastes which contain potassium chloride, potassium nitrate and strontium chloride.

Dr. Zahn Right. Sensitivity toothpastes desensitize the dental nerves directly. Potassium chloride, potassium nitrate and strontium chloride travel down the dentinal tubules to the dental roots and protect the dental nerves. Those toothpastes also contain sodium fluoride or aminfluoride which make teeth stronger and prevent caries. Ultra high fluoride mouthwashes and toothpastes can also help to reduce tooth sensitivity. After a few weeks of use the sensitivity usually disappears. However, foods which contain acids like citrus fruits, black tea, coffee or lemonade reduce the effect of sensitivity toothpastes. However, in some cases patients cannot get relief by brushing gently and using sensitivity toothpaste. Then we apply ultra high fluoride gel to the teeth, prescribe high fluoride mouthwashes and also apply a sealant to seal the cervical pores to protect the nerve-endings.

>>> ACTIVITY 1 >>>

Find these words or expressions in the text and add them to your vocabulary list!

hin und wieder, gelegentlich	Extraunterricht, Nachhilfe	fürchterlich, miserabel
Getränke zu sich nehmen	kalte Luft atmen	Anzeichen, Zeichen
ungeschützer Zahnnerv	die lateinische Bezeichnung	Zahnüberempfindlichkeit
Zahnfleischschwund	bedecken	die Zahnhälse
abhalten von	aus welchen Gründen ...	zurückweichen
zu viel Druck	stark abrasive Zahncremes	den Zahnschmelz zerstören

im Laufe von Wochen ...	das Zahnfleischgewebe	freilegen
es ist nicht gut	wie auch	weitere Gründe
richtig, ordentlich bürsten	Zahnbelag entsteht	mit der Zeit
verantwortlich für	abhalten von	den Druck vermindern
ordentliche Mundhygiene	Tipps zur Zahnhygiene	mindestens
Zahnseide verwenden	die Zahnoberflächen erreichen	Zeit aufwenden
weniger als	Zahncreme für empfindliche Zähne	welche enthalten
Kaliumchlorid und Kaliumnitrat	Strontiumchlorid	desensibilisieren
Zahnnerven	direkt	sich hinab bewegen, hinabwandern
Dentinkanälchen	Natriumfluorid	Karies verhindern
Fluorid-Mundspülung	die Empfindlichkeit lässt nach	Zitrusfrüchte
die Wirkung vermindern	in einigen Fällen	Erleichterung erfahren
sanft bürsten	Gel mit sehr hohem Fluoridgehalt auftragen	verschreiben
Mundspülung mit hohem Fluoridgehalt	Versiegelung, Dichtungsmittel auftragen	die Poren der Zahnwurzeln versiegeln

ACTIVITY 2

Look at the text and answer these questions!

1. Why can having sensitive teeth make life miserable?
2. Sensitive teeth is just a symptom. What is wrong with the teeth?
3. What is the Latin term for "tooth sensitivity"?
4. The tooth necks are normally covered by the gums. What function have the gums got?
5. How does the problem usually start?
6. Which reasons are the causes of receding gums/gum recession?
7. Is it good to use much pressure when brushing your teeth?
8. What will happen, if you do not brush your teeth properly?
9. What is responsible for the gingivitis and periodontal disease that cause gum recession?
10. Why should you also use dental floss after brushing your teeth?
11. What can patients do if they have sensitive teeth?

12. What effect do potassium chloride, potassium nitrate and strontium chloride have?
13. Why do most toothpastes contain sodium fluoride or aminfluoride?
14. Which foods reduce the effect of sensitivity toothpastes?
15. What can the dentist do, if sensitivity toothpastes do not work?

A joke a day keeps the doctor away!

(joke = Witz; a day = pro/jeden Tag; keep someone away = jemanden fernhalten)

ACTIVITY 3

Say it in English!

1. Empfindliche Zähne sind ein Anzeichen für ungeschützte Zahnnerven.
2. Das Zahnfleisch schützt die Zahnwurzeln.
3. Wenn das Zahnfleisch zurückweicht, verursachen die Zahnnerven Schmerzen.
4. Verringern Sie den Druck, wenn Sie Ihre Zähne putzen!
5. Wenn Sie zu viel Druck ausüben, tragen Sie das Zahnfleischgewebe ab und legen die Zahnnerven frei.
6. Wenn Sie Ihre Zähne nicht ordentlich putzen, entsteht Zahnbelag.
7. Mit der Zeit verhärtet sich der Zahnbelag und wird zu Zahnstein.
8. Die Bakterien im Zahnstein sind verantwortlich für Zahnfleischentzündung und Parodontitis.
9. Parodontitis ist eine Erkrankung des Zahnhalteapparats.
10. Sie sollten Ihre Zähne mindestens zwei Minuten lang putzen!
11. Wenn Sie empfindliche Zähne haben, rate ich Ihnen, eine nicht abrasive Zahncreme zu verwenden, die Kaliumchlorid, Kaliumnitrat und Natriumfluorid enthält.
12. Solche Zahncremes machen die Zahnnerven unempfindlich.
13. Mundspülungen, die Natriumfluorid oder Aminfluorid enthalten, können auch helfen, die Zahnempfindlichkeit zu vermindern.
14. Kaliumchlorid, Kaliumnitrat und Strontiumchlorid wandern die Zahnkanälchen hinab und schützen die Zahnnerven.
15. Nahrungsmittel, die Säuren enthalten, wie Zitrusfrüchte, schwarzer Tee, Kaffee oder Limonade, vermindern die Wirkung von Zahncremes für empfindliche Zähne.
16. Wenn solche Zahncremes nicht wirken, kann der Zahnarzt ein/e Dichtungsmittel/Versiegelung auftragen, welche/s die Poren der Zahnhälse versiegelt, um die Zahnnerven zu schützen.

ACTIVITY 4

Ask your partner these questions. Take turns!

1. What is the Latin term for "tooth sensitivity"?
2. How does the problem usually start?
3. For which reasons do gums recede?
4. If people use highly abrasive toothpastes and/or brush their teeth with too much pressure, what may happen to their teeth over years of brushing?
5. What is responsible for gingivitis and periodontal disease?
6. What would you advise patients suffering from sensitive teeth?
7. How do sensitivity toothpastes work?
8. Why do most toothpastes contain sodium and/or aminfluoride?
9. Which foods reduce the effect of sensitivity toothpastes?
10. What can be done, if patients cannot get relief by sensitivity toothpastes?

EXTENDED VOCABULARY LIST	
abrasion	Abrieb, Abnutzung, Abtragung (Zahnschmelz)
abrasive	abschleifend, abtragend, abrasiv
abrasives	Schmirgelstoffe, Putzkörper
agent	Agens, Mittel, Wirkstoff, Substanz, Beauftragter, Makler
apply	anwenden, benutzen, auftragen, auflegen
build up	aufbauen, entstehen, zunehmen, bebauen, sich verdichten
build in	einbauen
build on	anbauen
contain	enthalten, beinhalten
dentin	Dentin, Zahnbein
dentinal tubules	Dentinkanälchen
dentin caries	Dentinkaries

EXTENDED VOCABULARY LIST	
expose	entblößen, freilegen, aufdecken, bloßstellen, entlarven
floss **dental floss** **to floss**	Seide, Zahnseide Zahnseide Zahnseide verwenden
gentle	sanft, mild, wohlerzogen, gebildet
keep from	abhalten von
liquid	flüssig, flüssiger Stoff, Flüssigkeit
mildly abrasive	leicht abrasiv (mit wenigen und kleinen Schmirgelpartikeln)
miserable **It can make life miserable.**	fürchterlich, unglücklich, jämmerlich Es kann einem das Leben sauer machen.
once **once a week** **once more** **at once**	einmal, früher einmal einmal in der Woche noch einmal sofort, auf einmal
pores	Poren
potassium	Kalium
practice (US); practise (UK)	üben, praktizieren (Arzt), ausüben, trainieren
prescribe **in the prescribed time**	vorschreiben, verschreiben, verordnen fristgerecht
prescription **only available on prescription** **prescription charge**	Rezept, Verordnung, Verschreibung rezeptpflichtig Rezeptgebühr
prevent **to prevent someone from doing**	verhindern, verhüten, vermeiden jem. von etwas abhalten
prevention	Vorbeugung, Verhütung, Prävention
recede	sich entfernen, zurückweichen, zurücktreten
recession	Rückgang, Schwund, Rezession
relief	Erleichterung, Entlastung, Hilfe

EXTENDED VOCABULARY LIST	
root **dental root**	Wurzel, Grundlage, Ursache Zahnwurzel
seal	versiegeln, Siegel; Seehund (Zoologie)
sealant	Versiegelung, Dichtungsmittel, Abdichtungsmittel
sodium	Natrium
spend (spend/spent/spent)	verbrauchen, verbringen, ausgeben (Geld)
surface	Oberfläche, (Straßen)Belag
this time	dieses Mal
time **over time** **ahead of time** **all the time** **another time** **at times** **at all times** **by that time** **at the time being** **in time** **in no time** **on time** **to work against the time** **time is up** **Take your time!** **spare time** **waste of time**	Zeit mit der Zeit, nach einiger Zeit zu früh die ganze Zeit über ein andermal manchmal immer bis dahin zur Zeit rechtzeitig im Nu pünktlich unter Zeitdruck arbeiten die Zeit ist um Lassen Sie sich Zeit! Freizeit Zeitverschwendung
tissue	Gewebe, Papiertuch, Papiertaschentuch
tooth neck	Zahnhals
travel	Zahnhals
while **a long while ago** **for a while** **in a little while**	Weile vor langer Zeit eine Zeitlang in Kürze

For Your Information

As already mentioned, toothpastes containing abrasives like silica and calcium carbonate, can remove stains caused by tobacco, tea or coffee. Long-term use however, can destroy the dental enamel. That is why, children should never use toothpastes for adults. Such toothpastes contain higher fluoride concentrations and too many abrasives which are harmful for children's teeth. The abrasive effect of toothpastes is dependent on the "RDA", which stands for *Radioactive Dentin Abrasion*. The greater the RDA, the greater the abrasion. RDA up to 70 is mildly abrasive. RDA between 70 and 100 is moderately abrasive. RDA between 100 and 150 is strongly abrasive. RDA between 150 and 200 is very harmful. Below, you will find a list of toothpastes and their RDA. The list serves as an example and is not complete:

Brand (= Marke)	RDA
Kids Zahngel Logona	0
Kinder-Zahngel Weleda	0
Salviagalen F	0
Candida Peppermint	17–23
My first Colgate	29
Oral B sensitive	30
Elmex Sensitive	30
Signal Kids Zahngel	35
Theramed	40–43
Nenedent	40–50
Odol Med 3 Milchzahn	40–50
Parodontax	40
Parodontax F	43
Pearls & Dents	45
Mentadent C sensitive	43–51
Sensodyne F	54
Lacalut active	60

Brand (= Marke)	RDA
Mentadent C microgranuli	60
Rembrandt sensitive	64
Opalescence Zahncreme	66
Odol Med 3 40 plus	70
Perlweiß Raucherzahnweiß	70
Pepsodent Fresh Fluor	75
Meridol	75
Elmex	77
Aronal	80
Odol Med 3	96
Odol Med 3 samtweiß	96
Blend-a-med	100
Kräuter-Zahnpasta Logona	102
Colgate Sensation White	112
Smokers	167
Settima	180
Signal weiss	180

(source: www.zahnwissen.de)

(as already mentioned = wie bereits erwähnt; silica = Kieselerde; calcium carbonate = Schlämmkreide; remove stains = Flecken/Beläge entfernen; caused by = verursacht durch; long-term use = Dauergebrauch; that is why = deshalb; adults = Erwachsene; dependent on = abhängig von; stand for = steht für; up to = bis; mildly abrasive = mild abrasiv; moderately = mittelmäßig; exceed = überschreiten; it serves as = es dient als)

DENTAL DICTIONARY

dental	abrasion	Zahnabnutzung, Zahnabrieb
dental	abrasive strip	Separierstreifen
dental	abscess	Zahnabszess
dental	assistant	ZFA, zahnmedizinische/r Assistent/in
dental	administrative assistant chairside assistant	ZFA, Tätigkeitsbereich: Verwaltung ZFA, Tätigkeitsbereich: Stuhlassistenz
dental	alloy	Dentallegierung, Dentallot
dental	alveoli	Zahnalveolen, Zahnfächer
dental	alveolus	Zahnfach
dental	amalgam	Zahn-Amalgam
dental	anatomy	Zahnanatomie
dental	anomaly	Zahnmissbildung
dental	aplasia	Nichtanlage von Zähnen
dental	arch	Zahnbogen
dental	arch index	Zahnbogenindex
dental	arch width	Zahnbogenbreite
dental	articulation	dynamische Okklusion
dental	bill	Rechnung eines Zahnarztes
dental	bleaching	Zahntouchierung, Bleichen, Bleaching
dental	braces	Zahnspangen
dental	bridge	Brücke
dental	bur or drill	Zahnbohrer
dental	calculus tartar	Zahnstein
dental	calculus formation	Zahnsteinbildung
dental	care	zahnärztliche Versorgung, Zahnpflege
dental	caries Arrested dental caries	Zahnfäule, Zahnkaries Kariesstopp, Zahnkariesstillstand
dental	cast	Abdruck
dental	casting investment	Einbettmasse

dental	cavity	Zahnhöhle, Kavität
dental	cement	Zahnzement
dental	ceramics	Zahnkeramik
Dental	Chamber Federal Dental Chamber German Dental Association	Zahnärztekammer Bundeszahnärztekammer Bundeszahnärztekammer
dental	chart	Befundkarte mit Gebissschema, Zahnstatus, Zahnschema
dental	chisel	Zahnmeißel
dental	cleaning	Zahnreinigung
dental	conservation treatment	konservierende Zahnbehandlung
dental	cotton rolls	Watterollen aus Baumwolle
dental	crown	Zahnkrone
dental	cusp	Zahnhöcker
dental	cuticle	Zahnkutikula, Zahnoberhäutchen
dental	dam	Kofferdam, Spanngummi, Kofferdamtuch
dental	defects Grave dental defects	Zahndefekte gravierende Zahndefekte
dental	drill bit	Zahnbohrer-Einsatz
dental	elevator	Wurzelheber
dental	emergency service	zahnärztlicher Notfalldienst
dental	enamel	Zahnschmelz, Zahnemail
dental	eruption	Zahndurchbruch
dental	examination	zahnärztliche Untersuchung
dental	explorer	Zahnsonde
dental	examination results	Zahnbefund
dental	filling inlay	Zahnfüllung Inlay
dental	filling material	Zahnfüllstoff
dental	fistula	Zahnfistel, Zahnfleischfistel
dental	floss	Zahnseide
dental	fluorosis	Zahnfluorose

dental	focus	Zahnherd
dental	follicle	Zahnfollikel, Zahnsäckchen
dental	forceps	Extraktionszange
dental	fracture	Zahnfraktur
dental	gangrene	Zahngangrän
dental	granuloma	Zahngranulom
dental	grinding instrument	Zahnschleifinstrument
dental	hygienist	Dentalhygienikerin
dental	implant	Zahnimplantat
dental	impression	Zahnabdruck, Gebissabdruck
dental	infection	dentale Infektion
dental	inlay wax	Abruckwachs
dental	lamina	Zahnleiste
dental	membrane Alveolar dental membrane	Zahnmembrane
dental	mill	Zahnfräser
dental	mirror	Zahnspiegel
dental	neck Tooth neck Root caries	Zahnhals Zahnhals Zahnhalskaries, Zahnwurzelkaries
dental	nerve	Zahnnerv
dental	night guard mouth guard	Knirscherschiene, Zahnschutz
dental	panel doctor	Zahnarzt, der Mitglied einer Kassenzahnärztlichen Vereinigung ist
dental	periosteum Alveolar periosteum	Zahnwurzelhaut Zahnwurzelhaut
dental	pick scaler	Reinigungswerkzeuge: Zahnstocher, Scaler (to pick = stochern; to scale = schuppen, schaben)
dental	plaster	Gips
dental	plate	Zahnprothese
dental	porcelain	Zahnporzellan

dental	post	Wurzelstift
dental	office (US), surgery (UK)	Zahnarztpraxis
dental	preparation	Zahnpräparation
dental	products	Dentalerzeugnisse
dental	prosthesis Removable dental prosthesis Total dental prosthesis	Zahnprothese, Zahnersatz herausnehmbare Zahnprothese Totalprothese (= complete/denture/full)
dental	pulp Inflammation of the pulp	Pulpa Pulpitis
dental	retainer	festsitzende Retentionsspange, Haltespange
dental	root filling	Zahnwurzelfüllung
dental	root pin	Zahnstift
dental	scaler Double curved dental scaler pulse scaler sonic scaler Ultrasonic dental scaler Ultrasound dental scaler	Zahnreinigungsgerät, Zahnschaber doppelt gebogene Zahnreinigungs-kürette Impulsuntersetzer Schallscaler Ultraschallscaler Ultraschallscaler
Dental	science Doctor of Dental Science	Zahnheilkunde Doktor der Zahnheilkunde, Dr. med. dent., (Doctor medicine dentalis)
dental	sealant	Zahnversiegelung
dental	separator	Separiergummi
dental	stone	Modellgips
dental	surgery (UK) office (US) clinic (US/UK) center (US) centre (UK)	Zahnarztpraxis, Zahnklinik
dental	syringe	Injektionsspritze
dental	tape	Artikulationsfolie
dental	tips	Dentalspitzen

dental	trap	Absetzbehälter
dental	treatment chair	zahnärztlicher Behandlungsstuhl
dental	tweezers	Zahnpinzette
dental	utensils	Instrumentarium
dental	varnish lacquer	Zahnlack
dental	wax	Dentalwachs
dental	wax carver	Wachsschnitzer
dental	whitening strips	Zahnweißstreifen

VOCABULARY LIST

Deutsch	Englisch
abschleifen	grind down
abtöten	kill
abwechseln	take turns
Adresse	address
Akte, Datei	file
Anamnese	health history
Anfangsbuchstabe	initial
angreifen	attack
Angstgefühl(e)	anxiety
Anmeldung	reception
annehmen	accept
Antikoagulanz	anticoagulant
anwenden, auftragen	apply
Apotheke (US)	pharmacy
Arzt	physician
Arztbrief	medical letter
Assistent/in	assistant
ästhetisch	aesthetic
ausländisch	foreign
Atemproblem	difficulty breathing
auf der anderen Seite	across
auffallen, zeigen	show
aufsuchen, besuchen	visit
Ausdruck	expression
ausfüllen	complete, fill in
äußere Erscheinung	appearance
ausströmen	pour out
auswechseln	change
Backenzahn	back tooth
Bänder (med.)	ligaments
beeinträchtigen	affect
befestigen	fix
befreien von	to exempt from

Deutsch	Englisch
Befreiung	exemption
Befreiungsbescheinigung	certificate of exemption
Behandlungsstuhl	dental chair
beilegen	attach
Beißkraft	biting force
Belegschaft	staff
beschichtet	coated
Bestandteil	compound
bestimmt	definite/ly
bestimmt, sicher	certain
betreten, eintreten	enter
bewusst	aware
Bitte schön!	Here you are!
Blatt	sheet
Bleichen	bleaching
bluten	bleed
Bohren	drilling
Bohrer, bohren	drill
brechen, Bruch	fracture
charge, fee	Gebühr
Dämmerschlaf	half sleep
danach	afterwards
dauern, andauern	last
dennoch	all the same
Dentinkanälchen	dentinal tubules
Diamant	diamond
Diarrhoe, Diarrhö	diarrhoea
dringend	urgent/ly
Ecke	corner
Effekt, Wirkung	effect
eindrucksvoll	impressive
Eiter	pus
empfehlen	recommend
empfindlich (auf)	sensitive (to)
eng, fest, verspannt	tight

Deutsch	Englisch
entfernbar	removeable
entfernen	clean off
entfernen	remove
enthalten	contain
entscheiden für	decide on someth.
Entschluss, Beschluss	decision
entstehen	build up
entweder … oder	either … or
Entwicklung	development
entzünden	inflame
entzündet	inflamed
Entzündung	inflammation
Erdgeschoss	ground floor
erfahren, versiert	experienced
Erfahrung	experience
erfordern	require
erhalten	receive
Erleichterung	relief
ernähren von	feed on
ersetzen	replace
erwarten	expect
etwas dagegen haben	mind
extrahieren, ziehen	extract
Fachhochschule	college
fällig	due
falsche Zähne	false teeth
fangen, verstehen	catch
Farbe, Farbton Tönung	colour (UK) color (US)
Feind	enemy
fertig sein	done, finished
Flüssigkeit	liquid
Flur	corridor
fortschreiten (zu)	progress (to)
Fragebogen	questionnaire
freilegen	expose
Füllung, Plombe	filling

Deutsch	Englisch
Furcht, Angst	fear
ganz und gar nicht!	not at all!
Gebiss (auch künstl.)	dentures
Gebühr	charge
Gefahr	danger
gegenüber(liegend)	opposite
gelblich	yellowish
gelegen (geograf.)	located
gewachst	waxed
Giftstoff	toxin
gleich daneben	next to
Grund	reason
gruppenweise	in groups
Handzettel	flyer
härten, verhärten	harden
hässlich	ugly
häufig	frequent(ly)
Hauptmahlzeit	main meal
Hauptrolle	major role
Hauptunterschied	main difference
hell (auch geistig)	bright
herausziehen	pull out
Hier spricht … (Telefon)	This is …
hier: überweisen	refer
hilfreich	helpful
hinauf (im Gebäude)	upstairs
hinunter (im Gebäude)	downstairs
Hinweis	notice
höflich	polite/ly
Hülsenfrüchte	legumes
im Gegensatz (zu)	unlike
im Voraus	advance, (in)
Inzisor, Schneidezahn	incisor
Information (über)	information (on)
insbesondere	especially
Kalium	potassium

Deutsch	Englisch
Karies	caries
Kavität, Loch	cavity
Kellergeschoss	basement
Kiefer	jaw
Klang, Geräusch	sound
Klinik	clinic
Knirscherschiene	mouth guard, auch: night guard
Kohlenhydrat	carbohydrate
Kollege	colleague
krank	ill
Krankenversicherung	health insurance
Krankenversicherungsunternehmen	health insurance company
Krankheit (bestimmt)	disease
Krankheit (undefiniert)	illness
Kummer	trouble
Lächeln	smile
Latz	bib
lauten, lesen	read
Lehrbeauftragte/r	lecturer
leiden an	suffer from
leisten	afford
Lift (UK)	lift
Lift (US)	elevator
Liste	list
locker	loose
Lokalanästhetikum	local anaesthetic
mahlen	grind
Mahlzeit	meal
mechanisch	mechanical
Medikament, Medizin	medicine
Messer zum Schaben	scaler
Mineralwasser	mineral water

Deutsch	Englisch
mitkommen	come along
Mittel, Vermittler	agent
Modeschmuck	fashion jewellery
möglich	possible
Möglichkeit	possibility
Molar, Backenzahn	molar
momentan, zurzeit	presently
Mundspülung	mouthwash
Muskelverspannung	muscle tension
Na klar!	Sure!
Nachhilfe	tuition
nachlässig	careless, negligent
Nachname (UK/US)	surname
Nachname (US)	last name
Nachteil	disadvantage
Nackenschmerzen	neck pain
Nahrung	food
Narkose	anaesthesia
Narkosemittel	anaesthetic
Natriumfluorid	sodium fluoride
natürlich	natural
nehmen, annehmen	take
neutralisieren	neutralize
nicht stressig	stressless
Niederlassung, Ast	branch
nötig	necessary
Notiz	note
Oberfläche	surface
oberste Etage	top floor
ordentlich	proper
Ort (Ort)	location
paarweise	in pairs
Parodontitis, früher auch: Parodontose	periodontal disease
peinlich	embarrassing
persönlich, privat	personal
Phobie, krankhafte Angst	phobia
Plaque	plaque

Deutsch	Englisch
Polieren	polishing
polieren, Politur	polish
Pollen, Blütenstaub	pollen
Pore	pore
Porzellan, Keramik	porcelain
Praxis, Chirurgie, OP	surgery
Praxisgebühr	surgery charge
Problem	trouble
Prozent	per cent
Punkt, Fleck	spot
Rat (bezüglich)	advice (on)
raten	advise
Referat	lecture
regelmäßig	regular/ly
Rubin	ruby
sanft	gentle, gently
Saphir	sapphire
Säure	acid
schädigen	harm
schädlich	harmful
schlechter Atem	bad breath
schlucken	swallow
Schmerz	pain, ache
schmerzempfindlich	tender
Schmuck, Dekoration	decoration
Schneidezahn	front tooth, incisor
Schokoriegel	chocolate bar
schützen vor	protect from
Schwefel	sulphur
Schwellung	swelling
seelisch	mental
selbstsicher	confident
sich unterscheiden von	be different from
sicherlich	certainly
spezialangefertigt	custom made
spülen	rinse
Staatsangehörigkeit	nationality
stark, schlimm, streng	severe
Stoffwechsel	metabolism
stören, ärgern	bother
stressig	stressful
Stufen, Treppe	stairs
Tatsache	fact
Teil von	part of
Teilchen	particle
Thema	topic
Toilette (UK)	toilet
Toilette (UK/US)	lavatory
Toilette (US)	restroom
Toll! Großartig!	Great!
trotz, obwohl	in spite of
übel, schlecht, krank	sick
Übelkeit	sickness
überprüfen	check
überreichen	hand (to)
überschreiten	exceed
Überschrift	headline
Überweisung	referral
überzeugt	convinced
übrigens	by the way
übriggeblieben	left
Ultraschall	ultrasound
umwandeln in	turn into
ungefähr	approximate
unglücklich (über)	unhappy (about)
unschädlich	harmless
unterschreiben	sign
unzureichend, arm	poor
verbessern	improve
verbinden mit	bond to
vergleichen mit	compare to
verkleinern	minimize
verlegen, Schicht	shift
verlieren	lose
vermindern	reduce

Deutsch	Englisch
Verpflichtung	obligation
verschieden/e	several
verschönern, verbessern	enhance
Verschönerung	enhancement
verschreiben	prescribe
verschwinden	go away
Versteck	hiding-place
verstecken	hide
versteckt	hidden
verstoffwechseln	metabolize
Vertrag	contract
verursachen	cause
vielleicht	perhaps
Vollnarkose	general anaesthesia
Vordruck, bilden	form
vorlegen (Dokument)	produce, show, present
vorstellen	imagine
Wangen	cheeks
während	while
Wasserstoffperoxid	hydrogen peroxide
Weisheitszahn	wisdom tooth
wesentlich	essential
Wie kommt das?	How come?
winzig	tiny
wirksam	efficient
Zahlung	payment
Zahnen, Gebissform	dentition
Zahnfleisch	gum, auch: gums

Deutsch	Englisch
Zahnfleischgewebe	gum tissue
Zahnfleisch-schwund	gum recession, recession of the gingiva, gingival recession
Zahnhals	tooth neck
Zahnhygiene	dental hygiene oral hygiene
Zahnhygieniker/in	dental hygienist
Zahnkaugummi	dental gum
Zahnkrone	dental crown
Zahnmedizin	dentistry
Zahnnerv	dental nerve
Zahnpflege	dental care
Zahnschmelz, Emaille	enamel
Zahnschmerzen	toothache
Zahnseide	dental floss
Zahnstein	tartar, calculus
Zahntasche	periodontal pocket
Zahnverfall, Karies	tooth decay
Zahnwurzel	dental root
zerspringen	crack
zerstören	destroy
zuckerfrei	sugar-free
zum Leuchten bringen	brighten up
zurückweichen	recede
zusammenarbeiten	cooperate
Zustand	condition
Zustand, Staat	state
Zweifel	doubt

Englisch	Deutsch
accept	annehmen
acid	Säure
across	auf der anderen Seite
address	Adresse
in advance	im Voraus
advice (on)	Rat (bezüglich)
advise	raten
aesthetic	ästhetisch
affect	in Mitleidenschaft ziehen
afford	(sich) leisten
afterwards	danach
agent	Mittel, Vermittler
all the same	dennoch
anaesthetic	Narkosemittel
anaesthesia	Narkose
anticoagulant	Antikoagulanz
anxiety	Angstgefühl(e)
appearance	äußere Erscheinung
apply	anwenden, auftragen
approximate	ungefähr
assistant	Assistent/in
attach	beilegen
attack	angreifen
aware	bewusst
back tooth	Backenzahn
bad breath	schlechter Atem
basement	Kellergeschoss
bib	Latz
biting force	Beißkraft
bleaching	Bleichen
bleed	bluten
bond to	verbinden mit
bother	stören, ärgern
branch	Niederlassung, Ast
bright	hell, gescheit
brighten up	zum Leuchten bringen
build up	entstehen
by the way	Übrigens

Englisch	Deutsch
Bye!	Tschüss!
calculus, tartar	Zahnstein
carbohydrate	Kohlenhydrat
careless	nachlässig
caries	Karies
catch	fangen, verstehen
cause	verursachen
cavity	Kavität, Loch
certain	bestimmt, sicher
certainly	sicherlich
change	auswechseln, wechseln
charge	Gebühr, berechnen
check	überprüfen
chocolate bar	Schokoriegel
clean off	entfernen
clinic	(Poli-) Klinik
coated	beschichtet
colleague	Kollege
college	Fachhochschule
colour/color	Farbe (UK/US)
come along	mitkommen
compare to	vergleichen mit
complete	ausfüllen
compound	Bestandteil
condition	Zustand
confident	selbstsicher
contain	enthalten
contract	Vertrag
convinced	überzeugt
cooperate	zusammenarbeiten
corner	Ecke
corridor	Flur
crack	zerspringen
custom made	spezialangefertigt
danger	Gefahr
decide on someth.	entscheiden für
decision	Entschluss, Beschluss
decoration	Schmuck, Dekoration
definite/ly	(ganz) bestimmt

Englisch	Deutsch
dental care	Zahnpflege
dental chair	Behandlungsstuhl
dental crown	Zahnkrone
dental floss	Zahnseide
dental gum	Zahnkaugummi
dental hygiene	Zahnhygiene
dental hygienist	Zahnhygieniker/in
dental nerve	Zahnnerv
dental office	Zahnarztpraxis (US)
dental root	Zahnwurzel
dental surgery	Zahnarztpraxis (UK)
dentinal tubule	Dentinkanälchen
dentistry	Zahnmedizin
dentition	Zahnen, Gebissform
dentures	Gebiss (auch künstl.)
destroy	zerstören
development	Entwicklung
diamond	Diamant
diarrhoea	Diarrhoe, Diarrhö
different from	sich unterscheiden von
difficulty breathing	Atemproblem
disadvantage	Nachteil
disease	Krankheit (bestimmt)
done	fertig sein
doubt	Zweifel
downstairs	hinunter (im Gebäude)
drill	Bohrer, bohren
drilling	das Bohren
due	fällig
effect	Effekt, Wirkung
efficient	wirksam
either ... or	entweder ... oder
elevator	Lift (US)
embarrassing	peinlich
enamel	Zahnschmelz, Emaille
enemy	Feind
enter	betreten, eintreten
especially	insbesondere

Englisch	Deutsch
essential	wesentlich
examination	Untersuchung
exceed	überschreiten
exempt	befreien, befreit
exemption	Befreiung
expect	erwarten
experience	Erfahrung
experienced	erfahren, versiert
expose	freilegen
expression	Ausdruck
extract	extrahieren, ziehen
fact	Tatsache
false teeth	falsche Zähne
fashion jewellery	Modeschmuck
fear	Furcht, Angst
feed on	(sich) ernähren von
file	Akte, Datei
fill in	ausfüllen
filling	Füllung, Plombe
finished	fertig sein
fix	befestigen
flyer	Handzettel
food	Nahrung
foreign	ausländisch
form	Vordruck, bilden
fracture	brechen, Bruch
frequent(ly)	häufig
front tooth	Schneidezahn
general anaesthetic	Vollnarkose
gentle, gently	sanft
gingival recession	Zahnfleischschwund
go away	verschwinden
Great!	Toll! Großartig!
grind	mahlen
grind down	abschleifen
ground floor	Erdgeschoss
gum, auch: gums	Zahnfleisch
gum infection	Infektion des Zahnfleisches

Englisch	Deutsch
gum inflammation	Zahnfleischentzündung Gingivitis
gum line	Zahnfleischrand Zahnfleischlinie
gum recession	Zahnfleischschwund
gum tissue	Zahnfleischgewebe
half sleep	Dämmerschlaf
hand to someone	überreichen
harden	härten, verhärten
harm	schädigen
harmful	gefährlich, schädlich
harmless	ungefährlich, unschädlich
headline	Überschrift
health insurance	Krankenversicherung
health insurance company	Krankenversicherungs-unternehmen
helpful	hilfreich
Here you are!	Bitteschön!
hidden	versteckt
hide	verstecken
hiding-place	Versteck
How come?	Wie kommt das?
hydrogen peroxide	Wasserstoffoxid
ill	krank
illness	Krankheit (undefiniert)
imagine	vorstellen
impressive	eindrucksvoll
improve	verbessern
in groups	gruppenweise
in pairs	paarweise
in spite of	trotz, obwohl
incisor	Inzisor, Schneidezahn
inflame	entzünden
inflamed	entzündet
inflammation	Entzündung
information (on)	Information (über)
initial	Anfangsbuchstabe

Englisch	Deutsch
jaw	Kiefer
kill	abtöten
last	dauern, andauern
last name	Nachname (US)
lavatory	Toilette
lecture	Referat
lecturer	Lehrbeauftragte/r
left	übriggeblieben
legume	Hülsenfrüchte, Gemüse
lift	Lift (UK)
ligaments	Bänder (med.)
liquid	Flüssigkeit
list	Liste
local anaesthetic	Lokalanästhetikum
local anaesthesia	Lokalanästhesie
located	gelegen (geograf.)
location	Ort
loose	locker
lose	verlieren
main difference	Hauptunterschied
main meal	Hauptmahlzeit
major role	Hauptrolle
meal	Mahlzeit
mechanical	mechanisch
medical history	Anamnese
medical letter	Arztbrief
medicine	Medikament, Medizin
medication	Medikation
mental	seelisch
metabolise	verstoffwechseln
metabolism	Stoffwechsel
mind	etwas dagegen haben
mineral water	Mineralwasser
minimize	verkleinern
molar	Molar, Backenzahn
mouth guard	Knirscherschiene
mouthwash	Mundspülung
muscle tension	Muskelverspannung
nationality	Staatsangehörigkeit

Englisch	Deutsch
natural	natürlich
necessary	nötig
neck pain	Nackenschmerzen
negligent	nachlässig
neutralize	neutralisieren
next to	gleich daneben
Not at all!	Ganz und gar nicht!
note	Notiz
notice	Hinweis
opposite	gegenüber (-liegend)
pain	Schmerz
part of	Teil von
particle	Teilchen
payment	Zahlung
per cent	Prozent
perhaps	vielleicht
periodontal disease	Parodontitis
periodontal pocket	Zahntasche
personal	persönlich, privat
pharmacy	Apotheke (US)
phobia	Phobie, krankh. Angst
physician	Arzt
plaque	Plaque
polish	polieren, Politur
polishing	das Polieren
polite/ly	höflich
pollen	Pollen, Blütenstaub
poor	unzureichend, arm
porcelain	Porzellan, Keramik
pore	Pore
possibility	Möglichkeit
possible	möglich
potassium	Kalium
pour out	ausströmen
prescribe	verschreiben
presently	momentan, zurzeit
produce	vorlegen, produzieren
progress (to)	fortschreiten (zu)
proper	ordentlich

Englisch	Deutsch
protect from	schützen vor
pull out	herausziehen
pus	Eiter
questionnaire	Fragebogen
read	lauten, lesen
reason	Grund
recede	zurückweichen
receive	erhalten
reception	Anmeldung
recommend	empfehlen
reduce	vermindern
refer	hier: überweisen
referral	Überweisung
regular/ly	regelmäßig
relief	Erleichterung
remove	entfernen
removeable	entfernbar
replace	ersetzen
require	erfordern
restroom	Toilette (US)
rinse	spülen
ruby	Rubin
sapphire	Saphir
scaler	Messer zum Schaben
sensitive (to)	empfindlich (auf)
several	verschieden/e
severe	stark, schlimm, streng
sheet	Blatt
shift	verlegen, Schicht
show	auffallen, zeigen
sick	übel, schlecht, krank
sickness	Übelkeit
sign	unterschreiben
smile	Lächeln
sodium fluoride	Natriumfluorid
sound	Klang, Geräusch
spot	Punkt, Fleck
staff	Belegschaft, Mitarbeiter
stairs	Stufen, Treppe

Englisch	Deutsch
state	Zustand, Staat
stressful	stressig
stressless	nicht stressig
suffer from	leiden an
sulphur	Schwefel
Sure!	Na klar!
surface	Oberfläche
surgery	Praxis, Chirurgie, OP
surgery charge	Praxisgebühr
surname	Nachname
swallow	schlucken
swelling	Schwellung
take	nehmen, annehmen
take turns	abwechseln
tartar	Zahnstein
tender	schmerzempfindlich
This is ...	Hier spricht ... (Telefon)
tight	eng, fest, verspannt
tiny	winzig
tooth decay	Karies, Zahnverfall

Englisch	Deutsch
tooth neck	Zahnhals
toothache	Zahnschmerzen
top floor	oberste Etage
topic	Thema
toxin	Giftstoff
trouble	Problem
tuition	Nachhilfe
turn into	umwandeln in
ugly	hässlich
ultrasound	Ultraschall
unhappy (about)	unglücklich (über)
unlike	im Gegensatz (zu)
upstairs	hinauf (im Gebäude)
urgent/ly	dringend
visit	aufsuchen, besuchen
waxed	gewachst
while	während
wisdom tooth	Weisheitszahn
yellowish	gelblich